Making Ocean Policy

Also of Interest

Georges Bank: Past, Present, and Future of a Marine Environment, edited by Guy C. McLeod and John H. Prescott

Coastal Aquaculture Law and Policy: A Case Study of California, Gerald Bowden

International Politics and the Sea: The Case of Brazil, Michael A. Morris

Managing Ocean Resources: A Primer, edited by Robert L. Friedheim

† *Food from the Sea: The Economics and Politics of Ocean Fisheries,* Frederick W. Bell

The Politics of Pacific Island Fisheries, George Kent

† Available in hardcover and paperback.

Westview Special Studies in Ocean Science and Policy

Making Ocean Policy:
The Politics of Government Organization and Management
edited by Francis W. Hoole, Robert L. Friedheim,
and Timothy M. Hennessey

Written in response to the increasing interest in the making of ocean policy, this collection of original articles surveys the history of U.S. ocean policy, ocean policy advocacy, and the struggle within government to determine how best to develop and implement a sensible ocean policy. The increasing complexity of the issues, programs, and policies related to marine and coastal zone matters and the increasing number of government agencies and interest groups formed to deal with these matters reflect the growing awareness of their importance. But, reflect the editors, in an enormously complex world, where many interests are in conflict and where information is tentative and incomplete—yet often overwhelmingly abundant—there are few easy solutions to ocean policy problems.

Francis W. Hoole is associate professor of political science at Indiana University. **Robert L. Friedheim** is professor of international relations and associate director for marine policy studies, Institute for Marine and Coastal Studies, University of Southern California. **Timothy M. Hennessey** is professor of political science and associate of the Center for Ocean Management Studies and the International Center for Marine Resource Development at the University of Rhode Island.

Published under the auspices of the
Institute for Marine and Coastal Studies
University of Southern California
Los Angeles, California

and the
Center for Ocean Management Studies
University of Rhode Island
Kingston, Rhode Island

Making Ocean Policy

The Politics of Government Organization and Management

edited by Francis W. Hoole,
Robert L. Friedheim,
and Timothy M. Hennessey

Routledge
Taylor & Francis Group

LONDON AND NEW YORK

First published 1981 by Westview Press

Published 2018 by Routledge
52 Vanderbilt Avenue, New York, NY 10017
2 Park Square, Milton Park, Abingdon, Oxon OX14 4RN

Routledge is an imprint of the Taylor & Francis Group, an informa business

Library of Congress Cataloging in Publication Data
Main entry under title:
Making ocean policy.
   (Westview special studies in ocean science and policy)
   1. Oceanography and state – United States.   I. Hoole, Francis W.   II. Friedheim, Robert L.   III. Hennessey, Timothy M.   IV. Series.
GC58.M34   333.91'64   81-7418
ISBN 0-89158-966-X   AACR2

ISBN 13: 978-0-367-02245-7 (hbk)
ISBN 13: 978-0-367-17232-9 (pbk)

CONTENTS

ACKNOWLEDGEMENTS

This volume is being published under the auspices of the Institute for Marine and Coastal Studies of the University of Southern California and the Center for Ocean Management Studies of the University of Rhode Island.

We would like to thank Don Walsh, Director of the Institute for Marine and Coastal Studies, Donald L. Keach, Deputy Director of the Institute for Marine and Coastal Studies, and Virginia K. Tippie, Executive Director of the Center for Ocean Management Studies, for encouraging, sponsoring, and facilitating the creation and publication of this book. We gratefully acknowledge that financial support for volume development and production came from the Institute for Marine and Coastal Studies of the University of Southern California. We are also grateful for permission to use portions of four papers delivered at a conference sponsored by the Center for Ocean Management Studies of the University of Rhode Island. Marje Cappellari of the Institute for Marine and Coastal Studies did an excellent job of editing manuscripts and handling the production details for this publication. Her always cheerful suggestions and assistance greatly improved the final product. We would also like to thank Lynne C. Rienner, Associate Publisher, and Miriam Gilbert, Managing Editor, of Westview Press for their advice concerning publication matters. Finally, we would like to acknowledge the friendly cooperation of the authors of the chapters included here. The book would not have been possible without all of this support.

F.W.H.
R.L.F.
T.M.H.

INTRODUCTION

This volume is about the making of United States ocean policy. Its purpose is to provide perspective on contemporary U. S. ocean policy-making issues.

Such a volume is appropriate at this time for at least two reasons. First, ocean policies are becoming increasingly important in America. For example, policies regarding merchant ships, ports, continental shelf oil, nuclear reactor siting on coasts, building restrictions in the coastal zone, and maritime recreation now have significant economic as well as social implications, and they are becoming increasingly important as well as controversial. Second, as the importance of ocean policies increases, so does the importance of ocean policy-making processes. Accordingly, there is considerable contemporary interest in the organization and functioning of ocean policy-making processes.

The chapters presented here address themselves to a wide range of contemporary U. S. ocean policy-making issues. The volume contains chapters written by fourteen authors affiliated with ten different disciplinary departments in eleven universities. Ten of the fourteen chapters were written specifically for this book, while the others are extensive elaborations of papers presented originally in a set of conference proceedings. The volume contains chapters concerned with the history of the United States ocean policy, ocean policy-making advocacy, ocean policy-making analysis, and the study of ocean policy and policy-making.

History

Although the United States government has long been engaged in ocean affairs, it, nevertheless, seems appropriate, in retrospect, to identify activities that started in 1959 as the beginning of the era of a meaningful federal ocean program. In 1960, the Committee on Oceanography of the National Academy of Sciences released an influential study including recommendations for governmental action on ocean issues, and the Interagency Committee on Oceanography (ICO) was established as a permanent committee of the Federal Council for Science and Technology. During that same year the ICO attempted to identify and to catalog for the first time the various ocean activities of the federal government. Things developed rapidly after 1960. Among the highlights are the following: in 1963, a decade-long set of governmental goals was developed in regard to ocean activities; in 1966, the National Council on Marine Resources and Engineering Development was established, and the Sea Grant Program was initiated; the International Decade of Ocean Exploration was launched by the federal government in 1968; the National Oceanic and Atmospheric Administration was created in 1970; and in 1972 the Coastal Zone Management Act was passed by Congress. By the middle of the 1970s the pattern for federal ocean programs had been established, and state as well as local programs had been stimulated in most of the nation's coastal areas. By the late 1970s, the emphasis on the creation of new programs had changed to a concern with program implementation and evaluation. Clearly, the momentum for new governmental ocean activities which built up in the 1960s had dissipated by the end of the 1970s. As we enter the 1980s, we find a fresh look being taken at ocean programs and policy-making processes. Questions concerning the substance of policies as well as the organization of agencies and policy-making processes are being raised in numerous ways. The history of United States ocean policy-making is a complex, fascinating one. It is told in Chapter 1 by Dr. Robert B. Abel, a key participant in the development of the federal ocean program during the past two decades.

Advocacy

Many recent ocean policy initiatives call for the reorganization of the ocean activities of the federal government. These proposals are fueled by a widespread sense of the inadequacy of the current organization structure, which appears to encourage jurisdictional overlap and duplicate activities and includes no administrative focal point or coordinated advocacy of ocean programs at the highest levels of government. During the last eleven years there have been six major reorganization proposals: (1) the report of the Stratton Commission in 1969; (2) the report of the Ash council in 1971; (3) the Moore Proposal in 1976; (4) the Hollings proposal in 1976; (5) the report of President Carter's reorganization project in 1978; and (6) the report in 1979 of the National Advisory Committee on Oceans and Atmosphere. Each of these proposals called for a mechanism at the cabinet level to coordinate the ocean activities of the federal government, although there was no agreement on the exact nature of the coordinating mechanism. In Chapter 2, Robert E. Bowen of the University of Southern California presents overviews of the current federal organization structure and the six aforementioned proposals for reorganization. Key sections of the various proposals are included in this chapter.

Most advocates of reorganization call for the creation of a cabinet-level agency to coordinate federal ocean activities. Dr. Don Walsh of the University of Southern California disagrees. He feels that a reasonable solution lies somewhere between the current form of organization and the establishment of a centralized cabinet-level department. In Chapter 3, Dr. Walsh gives several reasons for the lack of a national ocean policy-making process, and he proposes rationalizing the ocean functions being carried out by the various federal agencies. He offers a "functional" proposal wherein the President is urged to use current authority to cut down on organizational overlap and duplicate programs. Dr. Walsh further proposes the establishment of a National Ocean Policy Council in the White House and an ocean policy directorate in the Office of Management and Budget.

Analysis

The last decade has seen the development of a small but expanding group of academic ocean policy analysts. They are affiliated primarily with cross-disciplinary, ocean-oriented programs at various American universities. These ocean policy scholars have found publication outlets in professional journals such as *Marine Technology Society Journal* and *Coastal Zone Management Journal,* edited volumes such as this one, and monographs. The current ocean policy literature can be characterized as being in a pretheoretical state with various issues serving as focuses. Several of the more salient issues are addressed in the analysis section of this volume by some of the more outstanding ocean policy scholars.

In Chapter 4, Dr. Stuart A. Ross of the University of Southern California focuses upon governmental reorganization problems, drawing from a wide variety of sources in organization theory and public administration. He observes that there is no single theory which satisfactorily indicates how to organize governmental activities. He does, however, discuss five useful orientations toward the problem: (1) the efficiency approach; (2) the political approach; (3) the humanistic approach; (4) the technology-environment approach; and (5) the social construction approach. Dr. Ross concludes that relying upon one approach or attempting a simple merger of the five approaches should be avoided in dealing with ocean agencies and activities.

The problem of how to organize the federal government to deal with ocean matters is, as emphasized earlier, currently of interest. An equally important and complex, but perhaps more subtle, organization issue is: How should the system of federal, state, and local governments be organized to deal with complex ocean-related problems? In Chapter 5, Professor Robert Warren of the University of Delaware sheds considerable light on this issue. Dr. Warren notes there is no agreement on the form, type, or scale of government organization for coastal zone activities, and he reports that there has been considerable experimentation in regard to these matters. A result is that a subtle modification in the balance of authority among local, state, and federal governments may be occurring, with control shifting toward the national level. Dr. Warren illustrates the complex issues involved by an analysis of the Coastal Zone Management Act of 1972. He argues that proponents of change in government structures or processes should be required to file a government impact statement which indicates possible effects and how compensation would be offered for undesirable impacts.

Professor Garry D. Brewer of Yale University argues in Chapter 6 that it is useful to conceive of ocean-related problems as having life cycles during which governments identify and define a problem, clarify possible solutions, adopt policies and programs, and implement, evaluate, modify, and finally terminate the policies and programs. Dr. Brewer views these activities from a decision-making perspective, and he discusses six major steps in the ocean decision-making process: (1) invention or initiation; (2) estimation; (3) selection; (4) implementation; (5) evaluation; and (6) termination. He also identifies apparent weaknesses in existing ocean decision-making processes.

The potential role of management science methodologies in ocean policy-making processes is of contemporary interest. Accordingly, in Chapter 7, I attempt to sort out this potential role. A discussion of the following types of policy-making processes is presented: (1) muddling-through; (2) modeling through; (3) comprehensive-integrating; and (4) unitary-rational-actor. A perspective on these processes is offered, and there is a discussion of twelve problems in utilizing the management science techniques. It is concluded that management science methodologies have a somewhat limited role in ocean policy-making processes.

The decision-theoretic orientation is one of the newer and more interesting approaches to the management sciences. One of the more promising versions of this approach is contained in the set of techniques that comes under the label of multi-attribute utility measurement

(MAUM). The essence of MAUM is the systematic collection and aggregation of subjective value judgments from policy-makers. Professor Peter C. Gardiner of the University of Southern California discusses a ten-step MAUM process in Chapter 8. Dr. Gardiner illustrates the use of MAUM by means of a coastal zone study in California. He also suggests ways in which multi-attribute utility measurement techniques can be used in typical coastal zone policy-making situations.

What is the economic value of the ocean to the United States? Professor Giulio Pontecorvo of Columbia University addresses this question in Chapter 9 in a report on a project of the Columbia University School of Business and the United States Department of Commerce. Professor Pontecorvo initially focuses on the development of a macro-level framework and a methodology for the measurement of the ocean sector. He then reports that the ocean sector comprised approximately three percent of the U.S. gross national product for 1972. Thus, the ocean sector appears to be about the same size as the agricultural sector, is larger than the mining sector, and is slightly smaller than the transportation and construction sectors of the nation's economy. Trade and governmental elements are the largest contributors to the size of the ocean sector.

The problems in designing and implementing an ocean program within the American federal system are examined in Chapter 10 by Professor Robert L. Bish of the University of Victoria and the University of Maryland. His analysis focuses on the design and implementation of the Coastal Energy Impact Program, which was created to handle local government fiscal impacts resulting from energy development. Dr. Bish examines how the diversity of state and local governments affects problem analysis, program design, legislation, implementation, and modification of a federally sponsored intergovernmental program. He highlights the role of organizational complexity and politics in effecting the design and implementation of technically designed programs in the American federal system.

Professor Willard Price of the University of the Pacific focuses on seaport management in Chapter 11. Dr. Price identifies the following key issues in seaport management: (1) the question of who should govern seaports; (2) port planning, including aspects of regional port planning, environment assessment, and coastal zone management; (3) the question of the need for an expanded federal port policy; and (4) the question of whether public enterprise finances are really independent. He argues that the concept of public enterprise offers a meaningful perspective for the study of seaport management. Within that perspective, Professor Price urges that further research focus on three aspects of the independence of seaport public enterprises from the authority of other governmental agencies: (1) governance, (2) management, and (3) finance.

Professor Timothy M. Hennessey of the University of Rhode Island focuses in Chapter 12 on the policy-making process used to implement the Fisheries Conservation and Management Act of 1976. Dr. Hennessey facilitates understanding of the complex fisheries management policy-making process by developing a set of hypotheses focusing on interactions between substantive issues, organizational structures, and information. He builds on the concepts of partisan mutual adjustment processes and disjointed-incremental decision-making patterns to begin the development of a positive model of fishery council decision-making processes.

Studying Ocean Policy and Policy-Making

What are the prospects for ocean policy research? Professor Lauriston R. King of Texas A&M University addresses this question in Chapter 13. Dr. King reviews four variations of research and policy guidance: (1) advisory reports; (2) short-term solicited projects; (3) systematic marine policy studies; and (4) historical and analytical case studies. He discusses several conditions inhibiting widespread support by federal agencies for academic ocean policy research. Professor King concludes with speculation about the prospects for such research. In general, he argues, the need will grow for ocean policy analysis as demands for governmental performance increase, and he feels the network of analytically sophisticated practitioners and scholars with ocean policy interests will increase in the future.

In what direction should academic ocean policy and policy-making research efforts be headed? Professor Robert L. Friedheim of the University of Southern California focuses upon this question in the final chapter. Dr. Friedheim reports on a National Science Foundation sponsored workshop which was convened to examine the state of the art in academic national ocean policy studies and to discuss future directions and funding for this research. He discusses eight major findings regarding the current state of national ocean policy studies, and he reports that discussions at the workshop generated eight basic suggestions regarding future research efforts. Professor Friedheim notes there was agreement that (1) it is important that national ocean policy studies continue to be conducted; (2) numerous funding agencies should support such studies; and (3) it is acceptable for the community of scholars to pursue several research focuses. Dr. Friedheim reports that a special emphasis on the study of the worldwide enclosure movement and its effects was called for by the conference participants.

An Orienting Perspective

I do not agree with everything in this volume, and some of the other contributors do not agree with each other. That is appropriate. I do respect the integrity of each writer, and I feel that the impersonal clash of ideas is necessary for progress in any field of inquiry. It is hoped that scholars, students, and practitioners concerned with ocean policy matters will find the varied ideas presented in this volume to be interesting and stimulating.

Francis W. Hoole
Bloomington, Indiana

HISTORY

CHAPTER 1
THE HISTORY OF THE UNITED STATES OCEAN POLICY PROGRAM

ROBERT B. ABEL*

In the Beginning

Selecting a starting date for the United States ocean policy program is largely an arbitrary judgment. Initial federal inquiry into the nature of America's land-sea interface began as far back as the administration of George Washington, and was, to some degree, a reflection of Washington's own personal interests. A surveyor himself, Washington commissioned a comprehensive survey of the coast of the emerging United States of America as it existed in his time.

Another milestone in the eighteenth century was Benjamin Franklin's initial delineation of the Gulf Stream in 1783.

In 1857, Matthew Fontaine Maury's analyses of ocean currents and his publication of the book *Physical Geography of the Seas*[1] were significant developments, as were the late nineteenth-century, large-scale marine biological enterprises undertaken by the Agassiz family.[2]

Given the paucity of significant oceanographic-related activity in the first hundred years of the country's history, it is apparent that efforts in the direction of a national ocean policy were minimal, a condition that existed through the beginning of the twentieth century.

However, immediately following World War I, the first glimmer of what could be interpreted as a national look at the ocean emerged. In the 1920s, Frank Schofield convened a U. S. Navy board to determine what contribution the Navy might make to a national effort in oceanography. This meeting resulted in the establishment of a pattern that has since been repeated with depressing regularity. The logic and far-sightedness of the committee's findings and advice notwithstanding, the recommended appropriations never materialized, plainly indicating lack of national

---

* Robert B. Abel is Vice President of the New Jersey Marine Sciences Consortium. He would like to note that since this chapter was written, the federal government has released the volume *Ocean Program 1976*, which describes federal agency activities in 1975 and updates somewhat the information on which this chapter was based.

[1] Matthew F. Maury, *Physical Geography of the Seas* (New York: Harper and Brothers, 1857).

[2] Louis Agassiz and his son Alexander directed a number of oceanographic expeditions aboard ships from the Fisheries Bureau and Coast and Geodetic Survey and founded the first marine geological laboratory in the United States in 1888.

priority. Thus, the principal contribution of the Schofield committee was to pave the way for another attempt.

This took place just a few years later, in 1927, when Dr. Frank Lillie, under the auspices of the National Academy of Sciences, undertook to formulate a national policy for the oceans and a program to advance United States ocean interests. Again, public response and support was minimal. However, at least one long-term result of that activity was the publication of the textbook, *Oceanography,* authored by Dr. Henry Bigelow, secretary of the committee.[3] The textbook remains a definitive one in the field.

If the government's failure to respond positively to these stimuli reflects some of the shortcomings of democracy—such as the need for an effective lobby or for a visible or audible constituency—then the sequel to the Academy's efforts must be considered emblematic of the free enterprise system. In response to the Academy's recommendations, the Rockefeller Foundation awarded grants totalling $16 million to establish oceanographic laboratories at Woods Hole and at the University of Washington, Seattle, and to aid the Bermuda Biological Station and Scripps Institution of Oceanography.

In fact, nearly all stimuli to oceanography in the period prior to World War II were provided by the private sector of the economy. Scripps Institution of Oceanography was founded by the Scripps family; the Rockefellers' significant contributions to oceanography have been cited; Captain G. Allan Hancock endowed substantial marine biology and oceanography activities at the University of Southern California; the Oceanographic Foundation at Yale University was endowed by Harry P. Bingham. The two individuals who, more probably than any others, stimulated oceanographic developments before World War II, William Beebe and Henry Bigelow, were both privately supported.

In short, throughout the period prior to World War II, the United States' national ocean program was essentially a privately supported program. This contrasts sharply with post-World War II developments, which led to the country's ocean program being widely characterized as governmental.

The advent of World War II provided a spectacular stimulus to oceanography in the United States and illustrates the dramatic impact that a single event can have upon national policy. The refinement of the principle of sonar resulted in a naval antisubmarine warfare program which constituted over half of the national budget for oceanography for

---

[3] An unusual byproduct of the deliberations was a textbook authored by the committee secretary, Dr. Henry Bigelow, entitled *Oceanography: Its Scope, Problems, and Economic Importance* (Washington: National Academy of Sciences, 1931).

the next three decades. Based on dollars and effort expended, it is tempting to mark the early 1940s as the beginning of the political history of the United States ocean program. However, I reject this thesis since the program, although large scale, was limited in that it dealt only with the acoustic properties of the ocean.

In one sense, my interpretation may be unfair. During this period there did exist, in addition to the aforementioned privately funded research, a program of fisheries research. However, it was small, primitive, and narrowly focused; but it was a modest federal ocean-related activity in a nonmilitary context.

The Years After World War II

The end of World War II allowed resumption of a number of oceanographic projects which had been subordinated to military effort.

It is important to note that although some projects had lain dormant during the war, World War II itself provided a significant stimulus to a number of widely diverse government oceanographic projects.

In the late 1940s, oceanographic research prospered modestly in a few institutions, most of which had been the original recipients of the Rockefeller grants: the Woods Hole Oceanographic Institution, Scripps Institution of Oceanography, and the University of Washington's Department of Oceanography. However, scientifically sound projects were just beginning at such institutions as the Universities of Miami and Rhode Island.

During this period of time, the normal research/education partnership was heavily skewed. Classroom enrollment was sparse, and the few oceanographers active during the late 1940s were more concerned with their research projects than with teaching. Indeed, although the first advanced degree in oceanography had been awarded in 1930, for the next three decades such awards were rare. As late as 1958, only 13 of the 2,780 science doctorates that had been awarded that year in the United States were related to oceanography—only one half of one percent.[4]

During the early 1950s, Scripps Institution graduated a small number of remarkable teacher/scientists and sent them out into the world as disciples. Dale Liepper started a program at Texas A&M University; Donald Pritchard, Wayne Burt, and Dayton Carrit inaugurated a similar effort at Johns Hopkins; and somewhat later John Knauss inherited and expanded Charles Fish's enterprise at the University of Rhode Island.

[4] Data estimated from *Doctoral Degrees in Science*, published annually by the National Academy of Sciences.

This activity may imply an upsurge of interest in oceanography, both in and out of government. Actually, the progress of the nation's ocean program—at least in the eyes of its practitioners—was disappointingly slow for the decade and a half following the end of the war. This probably resulted in part at least from the competition for resources during the transition from wartime to peacetime economy. In any case, progress was more sociological than material. For instance, the only major (depending on one's perspective) milestone in the 1940s was President Harry S. Truman's Proclamation 2667 of September 28, 1945, extending U. S. sovereignty over submerged resources on the outer continental shelf.[5] Furthermore, an attempt by the National Academy of Sciences to accelerate the ocean policy movement in 1949-51 proved ineffectual. However, as Vetter notes, "...in looking back, it is almost impossible to detect the effect of the 1951 study."[6] In fact, about the only discernible attempt at coordinating the country's ocean enterprises during the early 1950s occurred when marine scientists from the U.S. Navy Hydrographic Office (later renamed), the Navy's Bureau of Ships, and the Office of Naval Research began meeting more or less regularly to exchange ideas and information on their respective programs. They called themselves the "Informal Oceanographic Discussion Group."

In retrospect, this group assumes significance only when viewed against the backdrop of national lethargy. Given the low level of academic production and the low level of available resources, few dynamic individuals were attracted to the field from outside. It is probable, therefore, that the lack of significant progress was a result of the marine community's lack of size and, therefore, lack of clout.

In searching for causes of lack of progress, honesty compels me to make the observation that the ocean community was ineffective in building its case and presenting it to the public. In fact, one wit suggested that a good investment would be to give families of oceanographers fertility pills in order to develop a constituency in the long run.

Yet, even in the small ocean community, cooperation and coordinated enterprise was slow to develop among agencies and academic institutions alike. Edward Wenk calls attention to the rivalries among the major laboratories—a factor to which he assigns part of the blame for the lethargic movement toward a national ocean program.[7] Don Price

---

[5] Americans tend to castigate all of those nations who, earlier than we, declared 200-mile geographical limits, not realizing that these actions can be viewed merely as escalations of a philosophy first advanced by the United States.

[6] R. C. Vetter, "NASCO," in E. John Long (ed.), *Ocean Sciences,* p. 174.

[7] Edward Wenk, Jr., *The Politics of the Ocean* (Seattle: University of Washington Press, 1972), pp. 37-38.

concurs, stating that it appeared less difficult to "recognize the unity of nature than to develop unity in the oceanographic program."[8]

Perhaps the state of the dollar provides an even more dramatic index of malaise than does the number of degrees granted. In the same year (1958) that only one half of one percent of science doctorates were awarded in oceanography, less than $130 million of a national research and development budget of $15 billion was assigned to oceanography—again, one half of one percent.[9] Can this "poor relation" fraction be a coincidence? Viewed dynamically, the situation was no better; during the decade of the 1950s when the government R&D expenditures grew by a factor of five, oceanography support expanded by less than one half.[10]

Then, toward the end of the decade of the 1950s, the national ocean policy expanded, with academic and bureaucratic cacophonies that exceeded the sound of the surf itself.

Renaissance

The metamorphosis began relatively innocuously, and was characterized throughout its duration (some say it is still going on) by what amounted almost to collusion between academia and the federal bureaucracy (and perhaps, although unwittingly, by the Soviet Union which launched Sputnik, possibly the most effective catalyst in technological research and development in American history).

The first significant change occurred in 1956 when the supervisors of the members of the Informal Oceanographic Discussion Group, noticing a certain embryonic effectiveness of its activities, decided to organize on a somewhat more official, but still informal, basis. Operating without a charter, without officers, and without formal federal recognition, the Coordinating Committee on Oceanography (CCO), as it was called, nonetheless functioned for several years.

Opinions on the effectiveness of this group are mixed. Steven Anastasion notes that ". . .in spite of the informal character of its activities, the CCO in its early days influenced the status of Oceanography within the United States."[11]

---

[8] Don K. Price, *The Scientific Estate* (Cambridge, Mass.: Harvard University Press, 1965), pp. 211-212.

[9] Data taken from "National Oceanographic Program," published annually by the Interagency Committee on Oceanography.

[10] *Ibid.*

[11] Steven J. Anastasion, "Oceanography and Government," in Long (ed.), *Ocean Sciences*, p. 187.

R. C. Vetter has observed that "the meetings proved useful."[12] Vice Admiral John T. Hayward attributed the committee's viability to its informality and to rotation of responsibility and leadership among its member agencies.[13] Edward Wenk, on the other hand, is more conservative regarding the arrangement.[14]

The next significant development occurred in 1956. It originated with the CCO when officials of three of its member agencies, the Office of Naval Research, the Atomic Energy Commission, and the Bureau of Commercial Fisheries independently called upon the National Academy of Sciences to study the needs of the country relative to the ocean and to make recommendations. The academy responded enthusiastically. Its president, Dr. Detlev W. Bronk, was personally interested in the ocean, and had himself chaired the academy's Committee on Oceanography in 1949. In order to avoid the appearance of pro-oceanography biases, he appointed Dr. Harrison Brown, a distinguished geochemist from the California Institute of Technology and an "outsider" as chairman.[15] Members included mostly the giants of the oceanography profession. The executive secretary, Richard C. Vetter, has served the committee through its several incarnations and remains in that capacity today.

The USSR put Sputnik into orbit in October 1957, and in November the academy committee held its first meeting. It would be difficult to find a historical parallel in terms of chronology. As Wenk so colorfully puts it, "With one countdown, science was launched from the quiet and seclusion of the laboratory into the orbit of national policy."[16]

The effect on both the American public and its leadership was profound. President Dwight D. Eisenhower appointed a special presidential assistant for science and technology, later known as the President's Science Advisor. This person was selected by and from the members of a group of experts suddenly retitled the President's Science Advisory Committee (PSAC). From that point on, bureaucratic science proceeded with a rush. By 1962, science positions had been established at the policy level in the Department of Defense, the three branches of military service, and the Departments of Interior, Commerce, State, and

---

[12] Vetter, "NASCO," p. 175.

[13] U. S. Congress, Ocean Sciences and National Security, *Report of the Committee on Science and Astronautics*, 86th Congress, 2nd sess., July 1960 (Washington, D.C.: Government Printing Office, 1960), pp. 14-15.

[14] Wenk, *The Politics of the Ocean*, p. 50.

[15] Dr. Brown contributed to the group a strong sense of objectivity and thus enhanced its credibility to the public. This was a difficult situation. The perfect advisor is one who is close enough to the problem to understand it thoroughly, yet sufficiently far removed to remain completely objective. It is suggested that this person has never existed.

[16] Wenk, *The Politics of the Ocean*, p. 41.

Agriculture. All of these senior science executives were then assembled to form the Federal Council for Science and Technology (FCST), which was to complement PSAC within the Administration.

The FCST was established by Executive Order 10807 in March 1959. Included in the charter was a mandate to improve planning in science and technology, to promote closer cooperation among federal agencies, and to advise and assist the President in federal programs that affected more than one federal agency. One of the first problem areas considered by the council was oceanography.

Two months after the council was created, its chairman, James Killian, who was also Science Advisor to the President, announced the decision to form a temporary subcommittee to review the plans and programs of the various federal agencies charged with responsibilities that related to the ocean. In July 1959, the newly appointed first Assistant Secretary of the Navy for Research and Development, James H. Wakelin, Jr., was appointed chairman of that subcommittee. The subcommittee was not created without some tools. During that entire frenetic year of 1959, the importance of oceanography to the national concerns had been emphasized in various quarters. The PSAC had released, through the President, a special report on strengthening American science which cited the lack of attention being given to oceanography.[17] The Navy released its report on Ten Years of Oceanography (TENOC),[18] which was the first attempt by a federal agency to define the country's needs in terms of the ocean and to map a long-range course toward satisfying them. A month later, the National Academy of Science Committee on Oceanography (NASCO) released the first blockbuster in what was to be a series of nine reports issued over a two-year period entitled *Oceanography 1960-1970,* which identified national goals in terms of definite needs, suggested the processes by which these needs could be met, and proposed a set of milestones for the decade ahead.[19] The chapters were devoted to the areas of basic research, marine resources, defense, radioactivity, ships, engineering, history, and marine science in the United States. The last chapter of the report was completed in the spring of 1962. It identified national goals in terms of definite needs, processes by which these needs could be met, and a set of milestones for the decade ahead. The marketing process astounded even the authors of the report, who had

[17] U. S. President (Eisenhower), *Strengthening American Science,* Science Advisory Committee Report (Washington, D.C.: Government Printing Office, 1958).

[18] U. S. Navy *Ten Years of Oceanography* (Washington, D.C: Office of Naval Research, 1964).

[19] National Academy of Sciences, *Oceanography 1960-1970* (Washington, D.C.: National Academy of Sciences, 1961, 1962, 1963).

clearly not anticipated the level of response received from all quarters, particularly from the Congress. A Senate resolution, for instance, rather pointedly commended the report to the Administration.[20] Senator Claiborne Pell has called it "the first authoritative view of the great potential of the oceans."[21]

The National Academy of Sciences Committee on Oceanography (NASCO) was then invited to offer a special briefing to the Federal Council for Science and Technology, which referred the report to Secretary Wakelin's newly created ad hoc subcommittee. Thus, Dr. Wakelin and his colleagues, who included the directors of the principal agencies concerned with oceanography, did not want for de facto charter. Following a series of meetings and considerable review, the subcommittee generally endorsed the NASCO objectives and recommended its program. The subcommittee also recommended that it become a permanent committee under the FCST's charter.

On January 22, 1960, the council responded favorably to these recommendations by formally establishing the Interagency Committee on Oceanography as part of its own permanent structure.

Mobilizing for a National Effort

The Interagency Committee on Oceanography (ICO) will be given special attention here for a number of reasons. First, as we shall see, it represented a nearly new technique of federal administration; second, it provided the Federal Council and the President's Science Advisor with a means of confronting an increasingly questioning congress; third, both allied with and confronting the National Academy of Sciences Committee on Oceanography, it provided a means of mutual catalysis unprecedented in academic/federal agency relationships; fourth, it simply lasted longer than any of its predecessors or successors in what has seemed retrospectively to be a kaleidoscopic succession of federal extravaganzas; and fifth, 1959, the year of its birth, is generally regarded by observers to be the start of the U. S. "National Ocean Policy Program."

Organizationally, the ICO was an approximate shadow of its parent council; by 1962, its membership represented every sector of government concerned with the ocean. The first and most pressing need for the committee was to identify the oceanographic projects planned and/or underway in the federal bureaucracy and from these pieces form a U. S.

[20] As cited in Vetter, "NASCO," p. 179.
[21] Claiborne Pell, with Harold L. Goodwin, *Challenge of the Seven Seas* (New York: Morrow and Co., 1967) p. 169.

national ocean program. The steps contemplated were: identification, collection, consolidation, coordination, and unification. While the last never was achieved and perhaps never will be, ICO did make significant progress.

The first two steps were taken in 1960 when the ICO asked all federal agencies to nominate projects for inclusion in an ICO catalog and to submit budgetary information. With farsighted sagacity, the committee members agreed on which projects constituted a recognized ocean program. This designation provided the basis for a reasonable assessment of progress; the alternative—reformulating each year what should be part of the program—was recognized as bureaucratic chaos. With a baseline thus established, budgets could be aggregated as a measure of size to begin with, and as a measure of growth in succeeding years.

Having thus created the horse, i.e., program management, the committee could then apply its talents to constructing the cart, i.e., program planning. Nearly a year and a half prior to the start of a given fiscal year, the oceanographic aspirations of the agencies were collected by the committee staff and were sent to the appropriate panels; this was done by function, such as surveys, ships, etc. The panel members, as experts in their respective fields, would carefully and rigorously consider the merits of each agency's proposal, weighed against those of other agencies. Priorities thus developed at the operating level would then be presented to the full committee, perhaps further debated in terms of national and international policy implications, and ratified or modified. The program package was then presented to and defended before the Federal Council for Science and Technology.

It must be emphasized that it was this process of developing an annual ocean program on a grass roots planning basis that uniquely differentiated the ICO from both its predecessors and successors. It was, in fact, the only time in its history that the United States ocean program has been developed in such a fashion.

The process also enabled the committee to publish each year a national oceanographic program document which, in one form or another, endured for 15 years. At first this publication involved little more than a bookkeeping exercise in which all of the ocean project budgets of the federal agencies were aggregated and cross-accounted by function. As the committee members gained experience under the inspired and inspiring leadership of their chairman, Dr. Wakelin, they were able to exert increasing influence on the structure of the program, and thus the thrust of the annual report veered steadily from effect to cause.

By the same token, committee members worked progressively more closely as a team and were able to devote their attention to national issues to such an extent that in its waning years the committee had quietly assumed more centralized control over the country's ocean program than was generally realized.

Throughout the processes, the usefulness of the committee grew in a number of ways:

1. Based on work and findings of one of its ad hoc panels, the committee recommended to the federal council the creation of a National Oceanographic Data Center whose policies would be determined by several sponsoring agencies. The panel also designed the interagency agreement which, signed by the respective cabinet officers, served as the center's charter. The data center flourished and continues to function well.

2. Through its Panel on International Oceanography Affairs, the ICO became the focus of responsibility for nearly all U. S. government participation in international oceanography.

3. The ICO held the first national workshop on ocean instrumentation.

4. Making use of its multi-agency input, the ICO was able to design, prepare, and publish annual compendiums that were useful to all sectors of the ocean community. They included such documents as oceanographic ship operating schedules and oceanographic curricula in American universities. Special interpretive studies were made possible through the same devices, dealing with manpower, submersibles, bibliographies, and so forth.

5. Its growing cohesion enabled the ICO in 1963 to undertake a charter from the Federal Council for Science and Technology to attempt the development of a long-range plan for the country's ocean program.

This last assignment was particularly significant. Up to that time, the U. S. government had not constructed a consolidated long-range plan for its component agencies in any technological area. Based on its outstanding performance, the ICO was chosen to undertaken the prototype effort.

Here it is necessary to flash back briefly to 1960 when the House of Representatives Committee on Science and Astronautics published a volume entitled *Ocean Science and National Security*.[22] Although this work

[22] U. S. Congress, Committee on Science and Astronautics, *Ocean Sciences and National Security*, 86th Congress, 2nd sess., July 1960 (Washington, D.C.: Government Printing Office, 1960).

has since slipped into obscurity, it was, at this time, the clearest elucidation yet of the status of and outlook for marine science and technology. Its principal contribution, however, was to introduce to the federal ocean community its author, Edward Wenk, Jr., then director of the Library of Congress Science Policy Division.

Upon assuming his position with the FCST, Wenk was quick to see the potential of the ICO in realizing the goals expressed in *Ocean Sciences and National Security*. He served a useful role in gaining for the committee the inside track in council transactions. In fact, aided by Wenk's efforts and nurtured by President John F. Kennedy, who was personally interested in developing the ocean, the ICO and its program prospered. In 1963, the IOC produced *Oceanography—The Ten Years Ahead,* a statement of the government's long-range goals and the means by which they were to be pursued.[23] In its final form, this landmark document represented primarily the work of a single talented individual, Dr. Douglas L. Brooks, then vice president of the Travelers Research Center. Brooks was aided throughout by Dr. Arthur Maxwell of the Office of Naval Research, Dr. Harris Stewart of the Coast and Geodetic Survey, Dr. Wenk, and myself; but the synthesis was his alone.

As related by Dr. Wenk, the Interagency Committee on Oceanography was beginning to flower:

> In 1962-64, ICO continued to gain skill and momentum as a coordinating device. Its agenda focused on ways and means of strengthening the tools of research—providing greater ship capabilities, more space in shore based laboratories, and research funds to support an expanding enrollment of students and their faculties.[24]

At this point, however, a number of senators and congressmen, noting the success of ICO, began to explore various aspects of the question: If federal agency coordination is this effective in the absence of legislation, how much better might it be if it were given statutory underpinning? Furthermore, some congressmen were concerned with executive agency growth without controls. At that time, the congressional oversight network was chaotic, with each agency with marine-related responsibilities reporting to a different congressional committee.

Possibly the most authoritative description of congressional concern was articulated in a book authored by Senator Claiborne Pell (D-RI), Congress' strongest and most consistent supporter of a national ocean program, entitled *Challenge of the Seven Seas:*

---

[23] Federal Council for Science and Technology, *Oceanography—The Ten Years Ahead.* ICO pamphlet No. 10 (Washington, D.C.: Government Printing Office, June 1963).

[24] Wenk, *The Politics of the Ocean,* p. 78.

Of the 24 operating agencies, each has a statutory mission for which funds are appropriated by the Congress. Each is jealous of its prerogatives. Several are inveterate empire builders. The variation in competence in ocean matters is marked, ranging from extremely high to very low. Where competence is low, the reason usually is that the agency touches only lightly on sea affairs and may even resent the necessity of being involved at all. In a few of the old-line bureaus, ocean science and technology is a new and disturbing element, intruding on the calm, routine, well-ordered functions of many years. There is little cohesion among any except the agencies with direct ICO representation, and often there is not even mutuality of interest.[25]

The ICO, as a group, could have taken on the whole job of ocean development if it had been established under proper auspices and given a suitable supporting organization. The intelligence, the experience, and the devotion to the national interest were all present in the members and staff. What, then, was lacking? These factors:

1. A clear directive to take on the whole task.

2. Real power; the only kind of power that counts in Washington, in this kind of arrangement, is budget control, and ICO did not control the purse strings.

3. An adequate staff: the quality of the small, borrowed ICO staff was higher than such a jury-rig deserves, but the numbers were insufficient; a dozen such men with secretarial and clerical support would have made all the difference.

4. An administrative budget to support such a staff.

5. A different sponsor; a science-oriented hierarchy does not provide auspices for a comprehensive program that should contain much more than science.

6. A proper legislative base to enable the federal agencies to move into important fields that are not now part of their missions; this is discussed in more detail in the next subsection.[26]

Pell also provides the best analysis of the executive-legislative interactions existing at the time, showing that the twenty-seven agencies reported variously to fifteen congressional committees with many more subcommittees; that each agency reported to a different subcommittee of the House and Senate Appropriations Committees; that some individual

[25] Pell, *Challenge of the Seven Seas*, p. 252.
[26] *Ibid.*, pp. 263-264.

agencies actually reported to more than one committee; and that cooperative interagency projects could suffer if the committees happened to differ in their treatments.[27]

In so describing the complicated structure of executive-legislative interaction in ocean affairs, Pell was, in effect, voicing the concerns of a large number of distinguished legislators who had sought at various times to bring order into the system. Congressional interest in the ocean can be traced back to February 17, 1959, when a special subcommittee on oceanography was created in the House of Representatives under the chairmanship of George Miller of California. In the senate, Warren G. Magnuson, chairman of the Committee on Commerce, assumed similar jurisdiction.[28]

There were three types of legislative activity throughout that period. The first type related to the conventional oversight and appropriations process by which Congress exercises control over the operation of the individual executive agencies and thus over the *execution* of national policy. The second had to do with changes in the structures, missions, and authorities of the agencies by means of which Congress exercises its control over the nature of its agencies, and, thus, over the creation of national policy. The third and most dramatic type of legislation concerned the Congressional attempt to centralize and beef up the degree of clout that advocates of the ocean program had with the Administration.

In the first and third of these types of legislative processes, the executive and legislative branches found themselves in adversary relationships. As Pell has shown, a basic defect in the ICO technique was its (or FCST's, for that matter) failure to achieve budgetary integrity.[29] Thus, what could have been the committee's greatest strength—its coordinating function—paradoxically had the potential to become its greatest weakness unless all of the respective congressional committees and the component agencies which were tied to them achieved unanimity of approval in an interagency cooperative effort.

In the second instance, Congress can boast of a pretty fair record of achievement. In 1960, PL 86-409 removed geographical limitations upon the Coast and Geodetic Survey; in 1961, PL 87-396 removed limitations upon the Coast Guard's oceanographic program, and, in 1962 PL 86-626 did much the same for the U.S. Geological Survey. These actions were a mixed blessing, however; Wenk points out the Congressional hypocrisy in assigning a number of agencies the same authorities and responsibilities

---

[27] *Ibid.*, pp. 263-264.

[28] Don Price describes the early histories of the two committees whose initial efforts to organize oceanography in the executive branch were vetoed by President Kennedy in 1962. See Price, *The Scientific Estate*, chapter 4.

[29] Pell, *The Challenge of the Seven Seas*, p. 252.

as the Navy and then complaining of the duplication of effort in the executive branch.[30]

It was in the third legislative category that Congress found itself in the center ring. On the one hand, the ocean community was applying increased pressure for a national agency for the oceans (or a reasonable facsimile thereof); and, on the other hand, the Administration was maintaining that the status quo was optimal—that existing coordination was all that was needed.

Things changed in early 1965 when Senator Magnuson introduced S 944, which called for the creation of a national oceanographic council at the cabinet level. The next move—rather unexpected—came in the House of Representatives in June 1965 when Representative Paul Rogers, one of the staunchest supporters of a national ocean program, introduced HR 9064 in an attempt to establish a National Commission on Oceanography to study the potential for a national oceanographic program in depth at a top management level.

In the meantime, President Johnson's Science Advisor, Dr. Donald Hornig, appointed a special panel from the President's Science Advisory Committee to conduct its own study of the national oceanographic program, with particular reference to the plans and operations of executive agencies. The panel consisted of fifteen eminent scientists and engineers, not all of whom were oceanographers, and was chaired by Dr. Gordon MacDonald of Dartmouth College. Their report, released in July 1966, contained some 140 recommendations for new projects and improvements of existing programs. The highlight of the report was a recommendation for creation of a national agency for the ocean.[31]

As matters stood in mid-1966, the Administration was seriously reviewing oceanographic organization through the President's Science Advisory Committee; the Senate was still intent on passing Senator Magnuson's bill to create a cabinet-level council on oceanography; and the House was putting its best foot forward through Representative Rogers' bill for a blue ribbon oceanographic commission. Finally, on May 24, House and Senate conferees compromised on the Marine Resources and Engineering Act of 1966, featuring the cabinet-level council espoused by Magnuson, but making it temporary and adding the commission that Rogers advocated to ensure the continued top-level review of the ocean program.

This was landmark legislation, as viewed by the U.S. ocean community. As the new commission itself suggested, passage of the act and its signing by the President on June 17 as PL 89-454, expressed a conviction and defined a national purpose:

[30] Wenk, *The Politics of the Ocean*, p. 56.
[31] President's Science Advisory Committee, *Effective Use of the Sea* (Washington, D.C.: Government Printing Office, 1969), p. vi.

A conviction that the time had arrived for this country to give serious and systematic attention to our marine environment and to the potential resources of the oceans. A national determination to take the steps necessary to stimulate marine exploration, science, technology, and financial investment on a vastly augmented scale.[32]

National Council of Marine Resources and Engineering Development

The portion of the bill involving creation of the cabinet-level council was the first to be implemented. During mid-August 1966, Edward Wenk was brought back into the Administration as executive secretary of the council, and Vice President Hubert Humphrey, the council's chairman by statute, convened its first meeting. Viewed in retrospect, it would be difficult to identify a more dynamic duo than Vice President Humphrey and Dr. Wenk. The Vice President was brilliant and energetic in exercising leadership, perhaps because he had developed over the years an avid interest in the ocean. The national ocean program seemed made to order for him, and offered him a cause to which to devote his considerable energy.

Five months after the council was established, it prepared and the Vice President transmitted to the Congress via the President, a publication entitled *Marine Science Affairs—A Year of Transition,* in which the President told the Congress that:

We must launch a pilot program to assist the protein deficient countries of the world in increasing their capacity for using the fish resources of the seas (with particular reference to fish protein concentrate);

implement the Sea Grant College and Program Act to strengthen oceanographic engineering, expand applied research, and improve technical information activities;

accelerate studies to improve the collection, storage, retrieval, and dissemination of oceanographic data; expand ocean observation systems to improve near shore weather prediction services, and study ways to make more accurate long-range prediction of precipitation levels and drought conditions;

study the Chesapeake Bay to determine the effects of estuarine pollution on shellfish, health, recreation, and beauty, and to provide a basis for remedial measures;

[32] Commission on Marine Science, Engineering, and Resources, *Our Nation and the Sea* (Washington , D.C.: Government Printing Office, 1969), p. vi.

explore solid off-shore mineral deposits;

improve technology and engineering for work at great ocean depths; and equip a new Coast Guard ship to conduct oceanographic research in sub-Arctic waters.[33]

At the same time, the President transmitted to Congress his budget for a national ocean program, amounting to $462 million—considerably more than in former years. The council, in identifying its national ocean program, added a number of federal activities to what its predecessor, the Interagency Committee on Oceanography, had considered to be proper jurisdiction. Accordingly, the budget increase appeared larger than it actually was in terms of real dollars applied to real programs. Nonetheless, it did represent a significant step forward, demonstrating presidential initiative in the field of ocean policy.

In its executive structure, the council differed from the ICO, which was organized according to the basic functions of oceanography, ships, international programs, surveys, research, and education. The council placed more emphasis on national issues, creating committees on International Policy, Multiple Use of the Coastal Zone, Food from the Sea, Ocean Exploration, Environmental Services, and Marine Research, Education, and Facilities.

The activities of the Marine Science Council, presented through its five annual reports totalling nearly 1,000 pages, could easily be the subject of an entire book in themselves. Suffice to say that Congress assigned the council reasonable, if not lavish, resources which were spent in strenuous staff activities, a large number of outside consultant contacts, and a surprising amount of personal action at the council (i.e., cabinet) level which accomplished a number of aims:

1. Development of ocean science and technology in the United States was extensively publicized.

2. Fairly significant increases in funding were obtained, mostly through the appropriations process, during a five-year period in which the Bureau of the Budget (later named the Office of Management and Budget) never became particularly enthusiastic over either the council or its budgets.

3. Ocean causes were advocated at higher levels of government than had been previously attained.

This last point perhaps needs further analysis in retrospect, owing to the variability of the council's batting average. For instance, the fish

[33] President Lyndon Johnson, A Report to the Congress, *Marine Science Affairs—A Year of Transition* (Washington, D.C.: Government Printing Office, 1968), pp. 61-72.

protein concentrate objective stated in the Presidential message to Congress, never did get off the ground. To this day, its failure to develop into a useful vehicle for U. S. diplomacy abroad remains an enigma. The "collection, storage, retrieval, and dissemination of oceanographic data objective[34] was a legacy. The National Oceanographic Data Center had been designed and implemented by the Interagency Committee on Oceanography. Its greatest momentum was later reflected in the formation of the National Oceanic and Atmospheric Administration when oceanographic and meteorological systems were combined into the Environmental Data Service. The object of improving technology and engineering[35] materialized only partly. The failure of ocean engineering to develop as a first-rate national program still remains a mystery. Due to fits and starts and top management lethargy in a number of federal agencies, and despite a number of attempts on the part of industry to stir up action, the country still today lacks a healthy centralized progressive program in ocean engineering.

On the other hand, three programs which matured successfully owe much of their health to the council's nourishment. The first of these is coastal zone management. Wenk realized that the coastal zone was about to become the focus of frenzied activity as the aims and aspirations of fishermen, miners, conservationists, commercial cargo carriers, and people who wanted only to swim and sun were going to clash. Something had to be done about it. Wenk also foresaw intense confrontations between the environmentalists and the industrialists. Accordingly, while the first report of the marine science council ignored this issue, the council took up the matter of coastal zone management in its second year of activity when an appropriate committee was established; and the federal outlook was reflected in the second report of the council, *Marine Science Affairs—A Year of Plans and Progress,* published in March 1968. Chapter 5, entitled "Enhancing Benefits from the Coastal Zone," discusses the problem in all aspects, including engineering, water quality, health, conservation, recreation, and federal-state cooperation.[36] The report also included case studies of the Chesapeake Bay and the Seattle Harbor.

As is now well known, coastal zone management later became the focus of extremely strong and broad organizational development. Because the history of coastal zone management is so largely integral to the history of the National Oceanic and Atmospheric Administration, it will be discussed in detail later.

---

[34] *Ibid.*
[35] *Ibid.*
[36] President Lyndon Johnson, A Report to the Congress, *Marine Science Affairs—A Year of Plans and Progress* (Washington, D.C.: Government Printing Office, 1968), pp. 61-72.

The International Decade of Ocean Exploration

In 1968, as described in its January 1969 report,[37] the council decided to launch an International Decade of Ocean Exploration (IDOE). While Dr. Wenk had long espoused oceanography as an instrument of international diplomacy, the council's dealings with the subject were broad and general, involving such issues as cooperation in using marine resources, polar exploration, bilateral and multilateral foreign assistance, American commercial interests abroad, and so forth. The origin of the IDOE can be traced to the early spring of 1967 when Dr. Wenk, in his own words, began

> . . .to weld abstract scientific and policy goals into an explicit concept designed to interest and involve the entire domestic enterprise, to elicit policy support from both the Democratic and Republican Presidents, and to wash up on the shores of all coastal nations.[38]

This was an extremely broad effort which was envisioned later by the council as

> . . .a period of intensified collaborative planning among nations and expansion of exploration capabilities by individual nations, followed by execution of national and international programs of oceanic research and resource exploration so as to assemble far more comprehensive knowledge of the sea in a reasonably short time.[39]

The vehicle that was utilized was almost unprecedented in oceanography; the International Decade of Ocean Exploration was first announced in President Lyndon Johnson's State of the Union message on January 17, 1968.[40] The details came later in the President's special message to Congress on conservation on March 8, 1968. In the message, he stated:

> I have instructed the Secretary of State to consult with other nations on the steps that could be taken to launch an historic and unprecedented adventure—an International Decade of Ocean Exploration.[41]

In October 1968, Vice President Humphrey, the chairman of the council, assigned responsibility for planning, managing, and funding the

---

[37] President Lyndon Johnson, A Report to the Congress, *Marine Science Affairs—A Year of New Programs* (Washington, D.C.: Government Printing Office, 1969).

[38] Wenk, *Politics of the Ocean*, p. 213.

[39] President Lyndon Johnson, *Marine Science Affairs—A Year of Broadened Participation* (Washington, D.C.: Government Printing Office, 1969), p. 125.

[40] Wenk, *Politics of the Ocean*, p. 213.

[41] Johnson, *Marine Science Affairs—A Year of Plans and Progress*, p. iv.

program to the National Science Foundation. In doing so, he set out six goals. The first three concerned the preservation of the ocean environment, the improvement of environmental forecasting, and the expansion of seabed assessment activities. The remaining three involved hardware development, improved worldwide data exchange, and international sharing of responsibilities and costs for ocean exploration.

On the basis of these assignments, the IDOE staff, under the direction of Feenan Jennings, organized a program containing four major components: (1) environmental quality; (2) environmental forecasting; (3) seabed assessment; and (4) living resources. The fact that initiation of the IDOE was accompanied by $15 million of essentially new money gave the IDOE flexibility in the design of the program. The decision was made that the funds would be used only to study problems so large or so complex that they required the involvement of scientists from several institutions (or countries) and several scientific disciplines. Some of the individual projects involved as many as twenty-five scientists from twelve separate institutions. The IDOE became the big science complement to the many individual research projects which, until that time, represented the predominant means of conducting oceanographic research.

While IDOE never appeared to some of its early adherents to be sufficiently international, there is no question that a number of foreign countries participated effectively in the program. France, the Federal Republic of Germany, the USSR, Canada, India, Italy, Japan, and a few of the West African nations were among the earliest partners of the United States.

The National Sea Grant Program

A third achievement of the Marine Science Council was creation of the National Sea Grant Program. In this regard, it is built on the foundations laid before the council came into existence.

The National Sea Grant Colleges and Programs Act of 1966 (PL 89-688) appears to have been one of the last of the liberal products of an increasingly cautious Congress. The act was fashioned with sufficient clarity to provide guidance where guidance was needed, but with enough generality to permit free and imaginative program implementation.

The term "Sea Grant" was conceived by Dr. Athelstan Spilhaus in 1963, and upon his recommendation the concept was fashioned into law by Senator Claiborne Pell and Representative Paul Rogers.

The Sea Grant Program was created legislatively as Title II of the Marine Resources and Engineering Development Act of 1966. It was the

National Science Foundation which opened the program's doors in February 1967. The first project grants were awarded in the spring of that year; the first institutional grants were awarded in 1968; and the first Sea Grant Colleges were appointed in 1971.

Within NSF, Sea Grant was nurtured in an environment of experienced and enlightened science analysis and sponsorship. As a result, Sea Grant appeared destined for success. In fact, as viewed by other countries, Sea Grant had enormous potential. British analysts Brenda Horsfeld and P. B. Stone commented that:

> If the Sea Grant Program goes on to fulfill its objectives, America will soon have the most awesome capability in marine activities. The momentum provided by such a solidly based labour force will be irresistible, and, if there is wealth in the oceans, then the United States will get it.[42]

As it turned out, they need not have worried. First, Sea Grant was drawn into the administrative maelstrom surrounding the formation of the National Oceanic and Atmospheric Administration. Although Sea Grant was given a warm and hearty reception by the top management of the new Nixon administration, placement of the new and tiny staff among the in-house behemoths—the other NOAA bureaus—was akin to transferring a promising high school football team into the National Football League. Forward progress was maintained, however, under the personal support of the able NOAA administrator, Dr. Robert M. White. Sea Grant introduced several new or retailored concepts of federal science administration to the national ocean program:

1. A series of multi-project and interrelated programs

2. A concentration upon institutional problems

3. Maintenance of a balanced triple thrust: research, education, and marine advisory services

4. Establishment of status levels of program memberships: (a) individual projects; (b) coherent project programs; (c) institutional programs; and (d) Sea Grant colleges

5. Appointment of the bivalent Sea Grant program directors, who directed and coordinated the multi-campus programs

6. The establishment of the Sea Grant Directors Council, which gave the governed a voice in the government

---

[42] Brenda Horsfeld and P. B. Stone, *The Great Ocean Business* (New York: Coward-McCann, 1978), p. 174.

7. Formation of a strong, tough, experienced, and remarkably competent advisory panel

8. Encouragement of the development of social sciences such as economics, law, and sociology, and their fusion.

Although the program encountered budget difficulties throughout the 1970s, it faced its greatest peril in 1976, in terms of the formulation of what was to become the Sea Grant "Improvement" Act of 1977.

Although many of the portended legislative threats from this act, such as extraction of the Marine Advisory Services and administrative displacement, were warded off by sympathetic congressmen, the advisory panel's effectiveness was diluted by curtailing its term of service and changing its focus from technological to political.

Today, the program appears to be in good health, sponsoring fourteen Sea Grant colleges, an equal number of institutional arrangements of lesser status, and an aggregate of a thousand projects. Given appropriate fostering, the program may yet fulfill the British prophecy.

Commission on Marine Science, Engineering, and Resources (COMSER)

The Commission on Marine Sciences, Engineering, and Resources (COMSER), although formally established at the same time as the Marine Science Council, actually came into being somewhat later when President Johnson appointed a chairman and members early in 1967. According to the commission itself, it was:

> . . .asked to examine the nation's stake in development, utilization, and preservation of our marine environment; review all current and contemplated marine activities, and to assess their adequacies to achieve the national goals set forth in the act; and on the basis of its studies and assessments, to formulate a comprehensive, long-term national program for our marine affairs designed to meet present and future national needs in the most effective possible manner. . . . And, finally, . . to recommend a plan of government organization best adapted to the support of the program and to indicate the expensed cause.[43]

The commission was chaired by Dr. Julius Stratton, formerly president of the Massachusetts Institute of Technology, and at that time president of the Ford Foundation. The commission members represented (almost

[43] Commission on Marine Sciences, Engineering, and Resources, *Our Nation and the Sea*, p. vi.

equally) the federal agencies, industry, and academia, and were all distinguished in their fields. They divided themselves into task teams devoted to all of the major issues except that of government organization. In dealing with the latter, the commission organized itself as a committee of the whole. Staff was borrowed from a number of federal agencies, and nearly two years were devoted to painstaking and exhaustive study of national ocean policy.

Samuel Lawrence was detailed from the Office of Management and Budget to the post of director of the commission's staff, and Harold L. Goodwin was borrowed from the Sea Grant office to serve as chief editor of the report that would present the commission's findings.

Publication of the COMSER report, *Our Nation and the Sea*,[44] in January 1969 was given wide notice on the whole, but the section that attracted the most interest was the one relating to organization. Not surprisingly, the commission opted for a separate independent agency incorporating a large number of existing government functions and groups, including the Coast Guard, Data Center, Instrument Center, Environmental Sciences Services Administration, Bureau of Commercial Fisheries, Sea Grant Program, and Great Lakes Survey.

Thus, after ten years of cajoling, entreating, and haranguing, the ocean community received an endorsement of its aspirations for an ocean agency from a source which could hardly be ignored. Further, the report has remained the principal landmark from which progress has been dated and by which results have been assessed.

The creation of some sort of composite ocean agency was now inevitable.

COMSER, having delivered its four-volume report with literally hundreds of recommendations, as statutorily prescribed disbanded. As previously discussed, its most exciting recommendations concerned the creation of a national agency for the ocean. This idea had, after all, highlighted every report of every review committee ever chartered, but now it appeared that sufficient momentum for an agency had finally been acquired. At any rate, following some in-house maneuvering, on July 9, 1970, President Nixon issued Reorganization Plan No. 4 of 1970, creating the National Oceanic and Atmospheric Administration (NOAA).[45] The organization differed from the commission's concept in several ways, the major ones being that the Coast Guard was not included and that NOAA was housed with the Department of Commerce rather than being a separate agency.

The reorganization was partly successful in consolidating ocean activities of the executive agencies, by transferring from their parent

---

[44] *Ibid.*
[45] President's Reorganization Act No. 4, 1970. July 9, 1970.

agencies the Great Lakes Survey (Army Corps of Engineers), the National Weather Service (Commerce Department, Environmental Sciences Service Administration), Oceanographic Data Center (Navy), Oceanographic Instrumentation Center (Navy), Sea Grant Program (National Science Foundation), Coast and Geodetic Survey (Commerce Department, ESSA), ESSA Corps (Commerce Department), Environmental Satellite Service (Commerce Department, ESSA), and the Bureau of Commercial Fisheries (Interior Department).

After some delay, Dr. Robert M. White, Administrator of the Environmental Sciences Service Administration, was appointed administrator of the new organization. Howard Pollock, formerly congressman from Alaska, was appointed deputy administrator, and John Townsend was appointed associate administrator. Following a series of intensive evaluations, White organized his administration into: National Weather Service, National Mapping and Charting Service, National Environmental Satellite Service, National Marine Fisheries Service, Environmental Research Laboratories, Environmental Data Service, Environmental Satellite Service, Office of Sea Grant, and NOAA Corps. This structure indicates that Dr. White was emphasizing a focus on public service.

NOAA created programs early as a number of issues emerged that were of intense interest to the voting public; many of these were embodied in legislation, including the Ocean Dumping Act; the Fishery Conservation and Management Act; the Marine Mammal Protection Act; the Coastal Zone Management Act, the National Ocean Pollution Research and Development and Monitoring Planning Act; and the National Environmental Policy Act.

Coastal Zone Management

Undoubtedly, the most significant of all the programs to develop within NOAA was the Coastal Zone Management Program. The late 1960s were characterized by increasing public concern for the environment; much of this concern focused on the coastal zone—an arena in which waterborne commerce, recreation, commercial and sport fishing, and a sometimes unfriendly mother nature often competed. The Stratton Commission voiced its grave concern, stating that "something must be done."[46]

Immediately after the National Sea Grant Program was created, through it a number of university projects were mobilized to deal with technical and socio-economic coastal zone issues. By 1970, the Army Corps of Engineers had reoriented and expanded the scope of its urban

[46] Commission on Marine Science, Engineering, and Resources, *Our Nation and the Sea*, p. vi.

studies to assist state and local agencies in relating water problems in flood control, water supply, and waste water management to such problems as neighborhood renewal and transportation. Beginning in 1971, the Corps coordinated a study of Chesapeake Bay in which NOAA, the Geological Survey, NASA, and the National Science Foundation all participated.

In 1970, NOAA also inaugurated the long-term Marine Ecosystems Analysis (MESA) Program to "integrate and extend the agency's ability to describe, understand, and monitor the physical, chemical, and biological nature of marine environments; provide information and expertise required for effective management of marine areas and of their resources; and analyze the impact of natural or man-made alterations on marine ecosystems."[47]

Early in the 1970s, NOAA, the Department of the Interior, the Army, and the Environmental Pollution Agency collaborated on studies of the distribution and fate of pollutants in the coastal zone. In the Department of the Interior, the Geological Survey collected and analyzed data on dynamic process and water quality in 40 estuarine systems. The Coast Guard, in 1971, began a series of investigations of the sources of oil wastes on beaches. And, under a number of Congressional mandates, the EPA conducted many programs aimed at cleaning up the coastal zone environment.

At least twenty federal agencies became active in coastal zone planning and operating, and the number of projects aimed at preservation and optimal use of the coastal zone ran into the hundreds. Let it suffice to say that all of the previously mentioned agencies and many other agencies took their coastal zone assignments seriously, and in some instances developed programs based on the broadest and most liberal interpretations of their respective charters. In time, Wenk's phrase, "Multiple Use of the Coastal Zone"[48] more accurately characterized the interagency struggles for coastal zone *lebensraum.*

The most articulate advocate of coastal zone management was the Department of the Interior, which had conducted a number of studies during the 1960s aimed at reducing or resolving coastal zone problems. At the end of that decade, the Secretary of the Interior, Walter Hickel, funneled the results of all of these studies into well-prepared testimony before Congressional hearings that was highlighted by the statement that:

> The National Estuarine Pollution Study concludes that our estuarine
> areas are seriously polluted, and that the unwise use of the lands and

---

[47] The President, *The Federal Ocean Program* (Washington, D.C.: Government Printing Office, 1972), p. 14.

[48] Wenk, *Politics of the Ocean,* p. 230.

waters of our estuarine zones not only contributes to this pollution but is rapidly destroying valuable natural resources. While the statutory directive was to study the estuarine zones, the findings concluded that the management problems of our estuaries relate directly to the entire coastal zone and that any management system must deal with the Coastal Zone and its entirety.[49]

To that point in time, this was the most explicit, responsible, and comprehensive statement of the situation uttered at the policy level of government. It appeared certain that the Department of the Interior would be given authority over the country's coastal areas. Moreover, such assignment would dovetail with President Nixon's Land Use Program as an important component of it. Even after Land Use suffered the fate of a number of presidentially inspired programs during Watergate, the battle between the Departments of Commerce and the Interior for the coastal zone management program continued.

Passage of the Coastal Zone Management Act of 1972 allowed NOAA to achieve dominance. This actually represented the conclusion of only one battle in what has appeared, to some observers, to be a never-ending war for coastal zone jurisdiction between the Departments of Commerce and the Interior that really began with the placement of NOAA in the Department of Commerce.

The act itself is too complex to describe in depth, but its principal elements relate to:

1. Preservation and enhancement of the nation's coastal resources

2. Encouragement to the states to identify and carry out their respective responsibilities (e.g., management programs)

3. Cooperation between federal and state governments

4. Catalysis of cooperation among local and regional governments.[50]

Under the act, the Secretary of Commerce is authorized to award to state agencies: planning grants, development grants, administrative grants, and coastal energy impact fund grants.

Immediately upon receiving the legislative authority, the Secretary of Commerce created an Office of Coastal Zone Management within NOAA. Establishment of the program and the office was accomplished and

[49] U. S. Congress, Committee on Public Works, testimony given December 3, 1969; as quoted in J. M. Kreps, *U. S. Ocean Policy in the 1970s: Status and Issues* (Washington, D.C.: Government Printing Office, 1978), pp. iv-8.
[50] Commission on Marine Science, Engineering, and Resources, *Our Nation and the Sea*, p. 49.

greeted with fanfare. The program grew rapidly in size and importance. However, once the initial euphoria had subsided, reaction shifted. Some state governments began to weigh the merits of the money vs. the loss of autonomy and the prospect of confrontation with a formidable bureaucratic web. As the NOAA administrator himself admitted in 1976:

> The political climate for programs perceived as environmental in their thrust and those which involve additional governmental intervention and regulation is much harsher today than when the Coastal Zone Management Act was passed four years ago. States with coastal zone legislation on the books, at the time, are now fighting to prevent repeal of that legislation. In no case has preexisting state coastal legislation been strengthened. Without doubt, passing new state coastal legislation today is a much more difficult task than the framers of the Coastal Zone Management Act of 1972 envisioned.[51]

Apparently these framers had never bothered to ask the opinions of the state officials who would be saddled with the responsibilities. Finally, in 1978, the Secretary of Commerce, Juanita Kreps, admitted that the Coastal Zone Management Program had developed an intricate pattern of intergovernmental relationships; had redistributed power among federal, state, regional, and local governments; had caused conflicts among these governments and disagreements among constituencies; and had brought about delays, expense, and constraints among other federal agencies that had to work with state functionaries. She further stated that the highly touted interstate coordination was more advertised than real.[52]

Briefly, the elements common to approved Coastal Zone Management plans include:

1. Identification of the boundaries of the local coastal management zone

2. Definition and identification of land and water uses that have a significant impact on coastal waters and that will be permitted in the coastal zone

3. Designation and inventory of areas of particular environmental concern

4. Establishing priorities for uses, especially those of regional, statewide, or national significance, in particular areas of the coastal zone

---

[51] U. S. General Accounting Office, *Report to the Congress by the Comptroller of the United States, The Coastal Zone Management Program: An Uncertain Future* (Washington, D.C.: Government Printing Office, 1976), p. 101.

[52] Kreps, *Ocean Policy in the 1970s*, pp. iv-13.

5. Establishing procedures and organizational structure for managing the coastal zone

6. Identification of the authorities the state will use to exercise control over land and water uses

7. Description of the state organization structure that will operate the management program once it is approved

8. Definition of "beach" and planning processes dealing with access to public coastal areas

9. Planning of processes for dealing with energy facilities and their impacts, and for dealing with coastal erosion

A special arrangement has been made to handle the last element: the Coastal Energy Impact Fund. This fund, dispersed in loans, grants, and bond guarantees, assists states that are adversely affected by outer continental shelf development.

Today, the Coastal Zone Management Program affects most of the 50 states; a dozen and a half have had their plans accepted by NOAA. Acceptance by the local communities has been mixed, and perception of real benefits is the subject of discussion in a large number of intrastate conferences. Owing to the early demise of the land-use concept, however, coastal zone management is the only game in town.

And, Back to NOAA

While the birth and weaning of the Office of Coastal Zone Management had a strong and mainly salutary impact upon the National Oceanic and Atmospheric Administration, NOAA's responsibilities were clearly much broader than that particular function. NOAA's powerful interests and efforts in the atmosphere and space are beyond the scope of this treatment. The mapping and charting program expanded both technologically and geographically, covering all of the coasts of the United States and its territories. Storm prediction and evacuation projects became increasingly sophisticated and successful, particularly in preventing loss of life. In 1974, NOAA joined the march to seek out offshore fuel resources by instituting a marine geodetic program. The current NOAA fleet includes twenty-three ships of various types and sizes, engaged in mapping/charting, fisheries development, and oceanographic research and surveys.

Fisheries

The modern history (the last decade) of the U. S. commercial fisheries is largely within NOAA. The National Marine Fisheries Service, in between reorganizations, grew to its present size of 3,000 persons in twenty-six laboratories and the national headquarters. The history of the U. S. fisheries themselves is less impressive. This country, from its preeminent position in world fisheries after World War II, has descended to sixth position in total catch. At one point, personnel in the Office of Management and Budget were alleged (not confirmed) to have stated that the United States would be well off to abandon fisheries research and development completely and to depend on the results of research by foreign governments.

The Fisheries Service appears to be a perennial whipping boy, having changed names, functions, and organizations a dozen times since its origin a century ago. It was simply the Fish and Wildlife Service after World War II. In the late 1950s, the agency was split into the Bureau of Commercial Fisheries and the Bureau of Sport Fisheries and Wildlife. The two new agencies, however, reported to a Commissioner of Fish and Wildlife. Accordingly, cohesion remained in the system, even though an artificial management level appeared to have been inserted.

It was the creation of NOAA and its placement in the Department of Commerce that really played havoc with fisheries administration. Commercial fisheries became part of NOAA, and hence part of the Department of Commerce; sport fisheries remained in the Department of the Interior. While some stocks accommodate easily to categorization, others do not. Thus, such ludicrous delineations as the walrus belonging to the Department of the Interior and the seal to the Department of Commerce aided in muddying the waters of the national ocean program.

This organizational oddity did not result because of a lack of thorough review in the context of the national ocean program. Most of the early reviews convened by the President's Science Advisor gave careful attention to fisheries. The National Security Industrial Association created a special panel on fisheries in 1961. The President's Science Advisory Committee study, *Effective Use of the Sea*, devoted a special chapter to fisheries.[53] The Stratton Commission (COMSER) reviewed American fisheries administration in depth.[54] Myriad other advisory committees and panels, and professional and industrial associations and societies gave the government far more advice than it could ever hope to use.

[53] President's Science Advisory Committee, *Effective Use of the Sea*, pp. 5-15.
[54] Commission on Marine Science, Engineering, and Resources, *Our Nation and the Sea*, p. 40.

If the multiplicity of opinions demonstrated nothing else, it clearly showed how terribly the fisheries community of the United States is fractionated. Nonetheless, fisheries oceanography continued to receive close attention and funding during the 1960s and 1970s. Today, 3,700 plants in the United States employ more than 90,000 people, and another 160,000 fishermen operate in the harvesting end of the enterprise. The industry overall contributes in excess of $7 billion to the national economy in terms of food and various associated goods such as fertilizer, pet food, and so forth.

In terms of fisheries, the Fisheries Conservation and Management Act of 1976 is landmark legislation. This comprehensive and complex legislation encompassed many issues of concern to the fisheries community. Possibly the two most important were the establishment of a 200-mile economic zone and the creation of the regional fisheries council.

Traditionally a supporter of the 3-mile limit, the United States was reluctant to extend its boundaries. Like most naval powers, the United States worried about the reverse impact on freedom of the seas if all countries resorted to this action. An increasingly vocal fisheries community was able to demonstrate that our local waters were being fished out by Russian, Japanese, and other fishing fleets. The Fisheries Conservation and Management Act allows foreign fishermen to operate within the 200-mile limit only after United States fishing interests are satisfied and even then only if the foreign country has negotiated an agreement with the United States government.

The act also created eight civilian fisheries councils with more authority than the government normally delegates to such bodies. Briefly stated, the councils determine quotas by zone, by stock, and by season, and have powers of enforcement. They receive a share of the National Marine Fisheries Service budget for their operations. They are composed of public and private citizens representing local government, academia, and industry.

In recent years, probably spurred by the Fisheries Conservation and Management Act, emphasis appears to have shifted from research and development to management. Early in the 1970s, NOAA began work on a national fisheries plan. This was followed by work on a national aquaculture plan. In 1978, a Fisheries Development Task Force was formed to draft a new policy for the National Marine Fisheries Service, NOAA, and the government. The task force report, submitted on May 23, 1979, is one of the most detailed, comprehensive reports dealing with the problems of fisheries yet compiled.[55] Its highlights include:

---

[55] *National Aquaculture Plan,* a report of a Task Force to the Administrator of the National Oceanic and Atmospheric Administration, May 23, 1979.

1. Acknowledgement and explanation of the importance of the Fisheries Conservation and Management Act of 1976

2. A plea for partnership among industry and local, state, and federal governments

3. The suggestion that a national development policy will be implemented on a regional basis concentrated on nontraditional species

4. A proposal for new legislation for better funding support for fisheries development.

This report was translated into the Administration's policy and program statement on fisheries development during the same month.[56]

Industry and the Ocean

Industrial participation in the national ocean program began, and has progressed (as have most of the other components of the program) in fits and starts. Shortly after World War II, Secretary of Defense James Forrestal encouraged a number of industrial leaders to form the National Security Industrial Association (NSIA) for the purpose of advising the Department of Defense on plans and policies involving industry and military-industrial collaboration.

In 1963, a number of government and NSIA officials interested in the possibilities of ocean development organized a conference at the David Taylor Model Basin in Washington, D.C., to bring together industrial leaders interested in various aspects of ocean technology. Panels were formed on petroleum, fisheries, minerals, and navigation and salvage. To this structure was later added a panel on recreation. In a series of meetings over the next few years, the NSIA was active and useful in disseminating information about the government's oceanographic programs throughout industry for the benefit of those who would presumably be organizing and operating ocean-related businesses.

It had been evident from the beginning that ocean technology was nebulous, embryonic, and sufficiently expensive, and that it would take some sort of alliance between government and industry to make a go of most oceanography-related enterprises. Later in the 1960s, the NSIA formally organized an Ocean Science Technical Advisory Committee (OSTAC) to act as a permanent communication agent of the association. It should be stated, in this connection, that ocean development was the

---

[56] *Ibid.*

first area of technology in which the association made a deliberate effort to extend its interests and jursidictions beyond those of the Department of Defense. Acting through the Interagency Committee on Oceanography in the early and mid-1960s, the leaders of the NSIA committees became thoroughly familiar with the nature and activities of the various federal programs and their respective agencies. This interaction was influential in developing ocean-oriented programs in many large industries and in starting related new industries.

Typically, these early years of industrial ocean development were characterized by a number of studies and reviews accomplished by the Westinghouse Corporation, North American Aviation (for the Department of Defense), Battelle Memorial Institute (for the Coast and Geodetic Survey), the National Planning Association (for the National Council on Marine Resources and Engineering Development), the National Association of Manufacturers (*New Wealth from the Sea*), and a host of similar activities.[57]

In 1964, the National Academy of Sciences published a report on the economic benefits to be gained from ocean research.[58] For the first time, the principle of a cost-benefit ratio from ocean technology was introduced, and the academy proposed that returns from investments in certain areas of oceanography would triple within twenty years. This examination included fisheries, undersea mining, and weather and ocean storm forecasting.

Predictably, the Stratton Commission (COMSER) gave careful attention to the possibility of industrial development in the ocean.[59] The commissioners felt that the most likely industrial activities would be the development of continental shelf oil and gas; chemical extractions from sea water; mining of sand, gravel, and sulfur; and shrimp and tuna fishing. The commission observed that desalination was in an initial stage of growth, as was aquaculture. They felt that most segments of the fishing industry were already mature and were in many cases static or declining.

The commission suggested the possibility of future mining of placer minerals, development of oil and gas beyond the continental shelf, sub-bottom mining, deep-ocean recovery of manganese nodules, open-ocean aquaculture, and power generation from waves, currents, tides, and thermal effects. COMSER also pointed out that if American industry were to develop ocean resources successfully, it would need the closest possible

[57] *New Wealth from the Sea* (New York: National Association of Manufacturers, 1966).

[58] National Academy of Sciences, *Economic Benefits from Ocean Research,* Washington, D.C., 1964.

[59] Commission on Marine Science, Engineering, and Resources, *Our Nation and the Sea,* pp. 157-160.

government affiliation and strong support. Stating that "the key role of the marketplace is maintaining healthy, vigorous private activities in our economy," the commission emphasized the need for carefully designed government programs to accelerate the development of marine resource industries.[60] It further reinforced the earlier report of the President's Science Advisory Committee, *Effective Use of the Sea,* which had stated that "developments in undersea technology traditionally had resulted from industrial attempts to create new business opportunities in and under the ocean."[61]

Auxiliary but significant forces in influencing industrial development in the ocean were the professional and industrial societies such as the Ocean Council of the Institute of Electrical and Electronic Engineers (IEEE), the Marine Technology Society, and a number of engineering societies with ocean-related sections. The Marine Technology Society, formed in 1964, was particularly vocal in urging industrial investment in the ocean. By far the largest, most comprehensive, and most powerful industrial stimulus is the annual Offshore Technology Conference which currently attracts some 80,000 persons to exhibits of ocean-related (i.e., petroleum!) technology.

During the 1960s, one of the largest areas of industrial investment was in the design and fabrication of deep research vehicles. A number of American industries, encouraged by what appeared to be advancing federal initiatives in the ocean, produced, with or without government support, vehicles capable of observations at great depths. These included General Motors, Westinghouse, General Mills, North American, Reynolds Metals, and General Dynamics. In retrospect, it appears that this large-scale effort was partly misguided, first because of the government's failure to follow through on its plans, resulting in sharply reduced incentives; and second because the profit incentive suddenly became nebulous as it was discovered that the route from research to sales was more tortuous than expected.

As viewed by certain British observers, American industry's rush to get into the undersea industries had comic opera overtones.

> What did the big companies think they would get out of the ocean? None of them really knew, but they clung to the key word "exploration." An exploration vehicle seemed to most of them to be a suitable thing to begin with, a proper underwater device that could get a man right down there and let him get to grips with things. A research submarine had a comforting similarity to a space capsule. Above all, it was a master sales ploy. At Westinghouse, where the

---

[60] *Ibid.*
[61] President's Science Advisory Committee, *Effective Use of the Sea,* p. 20.

Deepstar submarines are made, they are quite clear about the psychology of it all. One dive equals one convert equals one customer. Lead time? Three years. "Yes sir, going down there really makes people think."[62]

As is often the case, disaster also influenced the course of industry in the ocean. The incident in which a hydrogen bomb was lost off Palomares, Spain, during an aircraft refuelling maneuver in 1966, and the destruction of the submarine *Thresher* in April 1963 advanced the rationale for further design and more sophisticated undersea vehicles.

Popular, semi-popular, and technical periodicals have also played roles in stimulating industry. In September 1965, the magazine *Missiles and Rockets* produced a special edition on oceanography which proposed a turning point in the ocean program at a time when aerospace development and employment appeared to be leveling off and even declining. *Fortune,* the *Wall Street Journal, Life, Time, Barrons, Business Week,* and *Scientific American* continually featured articles on productive ocean development, and numerous other publications occasionally took forays into this strange new field where opportunities were proclaimed to be unlimited.

However, considering all of the factors which led to and influenced the development of technology in the ocean and industrial participation in ocean technology, it must be concluded that the U. S. Navy was, by far, the greatest single stimulus. It was the Navy, particularly the Office of the Assistant Secretary for Research and Development, which stimulated the first activities of the National Security Industrial Association and which further chartered a number of industrial reviews of the ocean program. Their efforts were copied by several other agencies, including the Advanced Research Projects Agency, the National Aeronautics and Space Administration, the Department of Commerce's Coast and Geodetic Survey and Maritime Administration, and the Army Corps of Engineers. Nonetheless, the Navy effort probably exceeded the aggregate of the others.

Unfortunately, the Navy's efforts sagged in mid-flight. As Wenk observes:

> The aggregate ocean engineers' effort was dragging. The Navy was, by far, the best equipped to conduct ocean engineering, yet, Navy funding for ocean engineering remained level from 1967 to 1971. Considering inflation, this meant, in real dollars, a sharp decline.[63]

It should be noted that the Navy had principally spurred oceanographic research in the first place, starting in the Office of Naval Research in the

[62] Horsfeld and Stone, *The Great Ocean Business,* p. 178.
[63] Wenk, *The Politics of the Ocean,* p. 328.

late 1940s. The Navy was largely responsible for establishing ocean science programs at major universities such as Johns Hopkins, Texas A&M, Oregon State, and MIT. The Navy was influential in developing programs at the Scripps Institution, the University of Rhode Island, and the University of Miami, and certainly the Navy's contract program was the backbone of the development of the Woods Hole Oceanographic Institution. Further, the Navy, operating at least a half-dozen laboratories dedicated to ocean science and technology and about two dozen ships for oceanography and hydrography, certainly possessed by far the largest enterprise in the country and probably in the world. Reflecting the Navy's strong interest in oceanography, in 1961 the title of Hydrographer of the Navy, traditional among all the nations of the world, was changed to Oceanographer.

In its film, "Mission Oceanography," the Navy states that "The oceans were once barriers to progress; now they are secret pathways to our shores." As a final indicator of the Navy's interest in ocean science and development, it can be observed from the successive annual publications, including "The National Oceanographic Program,"[64] that the Navy's budget for this type of effort exceeded that of all of the other agencies until the late 1960s. It still occupies approximately one third of the national effort in spite of the growth of other ocean-directed agencies.

The early and middle 1970s represented a period of doldrums for industry; the malaise probably dates back to the major budget crunch in the late 1960s under the Nixon administration when not only did new funds originally promised for oceanography not materialize, but existing budgets were, in some cases, reduced. The large-scale grants and contracts anticipated by the private sector of the economy also failed to materialize. The early 1970s were characterized by a number of small companies going out of existence and large companies reducing or terminating their ocean efforts. Only the most dynamic companies (mostly petroleum-related) have maintained their momentum. Industrial planning prophecies are now conducted in a more conservative and exploratory mode.

Industry's most significant shot in the arm was the effect of its own initiative. In the middle 1960s, a small organization called the American Society for Oceanography was formed in Houston, Texas. It was initially financed out of a few gentlemen's back pockets, but the concept grew and spread to other cities. The organization grew to about 4,000 members and later partly merged with the Marine Technology Society. The remaining

[64] Federal Council for Science and Technology (Interagency Committee on Oceanography), *National Oceanographic Program* (Washington, D.C.: Government Printing Office; issued annually 1960-1967).

interests spun off to wander homeless until a few of the more active members decided to adopt stronger and sterner methods. Thus was formed the National Ocean Industries Association (NOIA), and Charles Matthews was brought in as its first (and present) director and president. An extraordinarily gifted organizer and entrepreneur, Matthews quickly pulled together the remnants and promoted the organization into its present membership of nearly 400 industries.

Today, the NOIA is possibly the strongest and most influential single force in the United States for ocean planning and policy. This notion may not be subscribed to by other observers of modern-day ocean science and technology; nonetheless, in considering objectively the decline of government coordination and organization, the weakening influence of academia on government affairs, and the general diffusion of responsibility of all sectors of ocean activity, it appears that NOIA is the most cohesive force in planning, coordinating, executing, and controlling ocean activities. Reference to congressional literature will show many instances in which testimony by this association has been influential in determining legislative development. NOIA's current status enables this description of a somewhat confusing alternation of progress and decline in industrial activities to end on a hopeful note.

Events of the 1970s

To the ocean community, establishment of the National Oceanic and Atmospheric Administration appeared to usher in the decade of the seventies on an optimistic note. In retrospect, it would have been difficult for the officials of NOAA to have fulfilled all expectations, given the severe constraints encountered at every level of administration. In most ways, this decade has been distinguished by progress; in other ways, by frustration.

1. Creation of NOAA was to have centralized ocean planning and management, and reduced fragmentation of authority. Actually, the number of federal agencies sponsoring and/or carrying out ocean projects has increased from 17 in 1970 to 38. Furthermore, Coopers and Lybrand report that in 1975 there were 54 public and quasi-public organizations involved in fisheries alone.[65] And, NOAA, which was expected to assume

[65] Kirk Juanda, *Responsibilities, Activities, Authorizations, of Federal Agencies Having Bearing on U. S. Marine Fisheries, both Commercial and Recreational*, and Related Industries. A report issued by Coopers & Lybrand (Washington, D.C.) September 1975. NTIS: PB 254250.

and exercise leadership in the federal ocean program, commands only 8 percent of the federal ocean budget.

2. Congress passed nearly a dozen major ocean-related laws during this decade, spreading responsibility for their implementation throughout the government.[66] Congress then castigated the administration for waste and duplication of effort.

3. Since World War II, the ocean and ocean programs have assumed increasingly more important roles (and larger budgets) in the affairs of the country. Yet the focus of responsibility is more blurred today than at any time in recent history.

4. A major objective of the Stratton Commission and a primary intent of its report was to achieve a degree of stability in the federal administration of ocean programs. In fact, the present decade has been marked by administrative fluidity as reorganization has become almost a fetish in some departments.

5. The common property of ocean legislation is the reporting function; the intent of Congress is clearly to disseminate the maximum amount of data and information concerning ocean affairs to the public. Yet, the *National Oceanographic Program* begun by the ICO twenty years ago, which became, successively, the *Marine Science Affairs* and *Federal Ocean Program* series, has now lapsed, apparently owing to Administration apathy.[67]

This section will attempt to chronicle in some reasonable sequence these apparently inconsistent actions, many of which depend on the capricious nature of politics.

The departure of Hubert Humphrey from the executive branch of government in 1969 presaged a decline in the fortunes of the ocean community. His successor, Vice President Spiro T. Agnew, evidenced

---

[66] The major laws included: Federal Water Pollution Act Amendments of 1972; Marine Mammal Protection Act; Marine Protection Research and Sanctuaries Act of 1972; Coastal Zone Management Act of 1972, and Amendments of 1976; Deep Water Ports Act of 1974; Fishery Conservation and Management Act of 1976; National Ocean Pollution Research and Monitoring Planning Act of 1978; and Outer Continental Shelf Lands Act Amendments of 1978.

[67] It is suggested that, strictly speaking, discontinuance is illegal; Section 7(a) of the Marine Resources and Engineering Act of 1977 (PL 89 454), still operative, demands an annual accounting. See President Johnson's *Marine Science Affairs—A Year of Broadened Participation,* Appendix B2.

little interest in either the ocean program or its management. The result was the Marine Science Council's demise in 1971, ushering in the dark ages of marine science coordination in government.

A none-too-robust Phoenix arose from the council's ashes shortly thereafter, when a number of responsible agency heads, by now addicted to at least minimal coordination, formed the Interagency Committee on Marine Science and Engineering (ICMS). The charter for this group was nebulous, but the ICMSE was exceptionally well staffed by Capt. Steven N. Anastasion, USN (Ret) and William Windom, who had been prominently identified with the Interagency Committee on Oceanography (ICO) a decade previously.

The ICMSE is difficult to understand; although superficially related to it, it was clearly not intended to be the same as the Interagency Committee on Oceanography. The environment in which it was created was not the one that ICO had found. The ICO was a pioneer in a wilderness of people looking for somebody to take an interest in the oceans. ICMSE, on the other hand, was born into a world in which *everybody* had an interest in the ocean and cast wary eyes on anybody or any committee that might tend to usurp, even loosely, some of the established prerogatives.

Its objectives were more ocean research and development policy than pragmatic program solutions, as was required in the days of ICO (which had inherited a formless program). Consequently, the spectre of protectionism manifested itself far more within ICMSE than it had ever had to in ICO. This does not mean that there was a lack of cooperative attitude in ICMSE. Certainly NOAA, Navy, and NSF relationships were excellent because of the personal relationships of their leaders. On the other hand, other members strongly opposed ICMSE considerations of activities which they believed to be their sole prerogatives. Moreoever, consideration of anything but research and development by ICMSE was thought to be out of order by a few of its members. Even so, within this environment, a number of useful things were accomplished, and a recognized form of interagency communications was established.

ICMSE's biggest problem was the failure to convey its message to the necessary influential White House people through its parent committee.[68]

In addition to the "communications upward" problem, another symptom of the managerial malaise affecting the national ocean program of this era was the apparent division of responsibility between effecting coordination and reporting of it. The Office of Science and Technology (OST) in the Executive Office of the President assumed authority over the ocean program when the Marine Sciences Council disbanded, but that office, in turn, was eliminated shortly after in 1973. During that period,

[68] Personal communication from Steven Anastasion.

the ocean program was without an obvious advocate in the OST, and this lack of influence and high-level encouragement was reflected in the presentation level at ICMSE meetings as well as in the previously mentioned communications barrier. Part of the motivation for Dr. White and his colleagues to organize the ICMSE related to their recognition of the Administration's responsibility for keeping the Congress and electorate informed concerning the status of the ocean program. The question was: Who should be the author of record?

After some deliberation, the ICMSE came up with *The Federal Ocean Program*,[69] a highly respectable heir to the legacy of the *National Oceanographic Program* and *Marine Science Affairs* series. This report, like its predecessor, ultimately represented a report from the President to the Congress, but the conveyance thereof was less clear cut.

In 1973, because of the absence of a science functionary at the White House level, responsibility for supervision of the remnants of the ocean program shifted to the director of the National Science Foundation in his ancillary capacity as the President's Science Advisor. NSF's Science and Technology Policy Office, however, proved unequal to the task of leadership of so many diversely oriented agencies in such a complicated program.

As a result, actual program coordination was left in the hands of the ICMSE, while the reporting function still required the endorsement of the director of the National Science Foundation prior to publication. Evolution in such a dichotomous situation was, naturally, in the direction of procrastination; thus, by 1976, the annual report was delayed by a year and a half; by 1977, two annual issues were backed up awaiting approval; and, by 1979, the whole idea had apparently been given up as a waste of time. At present, therefore, the last annual status report of the *National Ocean Program* is four years old, and after a decade and a half of thoughtful and comprehensive enunciation of national aspirations and activities, the federal agency complex, twice as large and fractionated as a decade ago with greatly expanded responsibilities assigned by Congress, has gone silent. Thus, the thread of continuity started in a remarkably simple fashion by the ICO in 1960 has now been lost.

A frustrated 94th Congress, seeking to restore momentum to science management in the executive branch, enacted the National Science and Technology Policy Organization and Priorities Act of 1976. For the first time, science leadership had statutory underpinning; it restored an operating Office of Science and Technology Policy and revived the Federal Coordinating Council for Science, Engineering, and Technology (FCCSET).

---

[69] *The Federal Ocean Program* (Washington, D.C.: Government Printing Office); issued annually 1974-1976.

At that time, according to Secretary of Commerce Juanita Kreps, Dr. Frank Press, the President's Science Advisor (who was a prominent geologist) chose to combine the ICMSE and the Interagency Committee on Atmospheric Sciences (which had survived in its original form for 17 years) into a Committee on Atmosphere and Oceans (CAO), under his personal chairmanship.[70] This is disputed, however, by the staff of the President's Science Advisor, who felt that from its inception, the committee (CAO) was always subordinated to the council (FCCSET).[71]

This move should have been salutary from the viewpoint of the ocean community; the ICMSE, although a good vehicle for interagency communications, had been less effective in influencing ocean policy at the White House level. Responsible oversight at that level has always been the fundamental goal of ocean program leaders who ascribe most of the blame for the slow pace of the national ocean program to lack of a high-placed advocate.

This sort of reasoning is questionable. For instance, the Department of Commerce, in *U. S. Ocean Policy in the 1970s,* identifies a number of sand traps associated with placement of responsibility for ocean policy at the President's doorstep, relating mainly to competition for his attention with more critical issues.[72] The same sort of argument is also advanced to rebut the proposition for a national agency for the oceans. Yet the nagging thought persists: Why is it that, without exception and regardless of its composition, every group that has been empaneled to review the National Ocean Program has eventually recommended the independent agency? Recalling their examinations, the majority of the members of the NSIA committee, PSAC, and COMSER were not ocean-oriented prior to their appointments to the committees; yet they all became advocates, and their advice continued to be promotional. It is quite possible, in fact, that the process of giving and taking advice has been of greater significance to the evolution of the national ocean program than to the development of other enterprises of this nature.

Advisory Activities

While the advisory committee is hardly a modern-day phenomenon, it has attained its present degree of importance to government only since World War II. Don Price discusses the crucial function of advisory mechanisms in the executive/legislative interactive strategies of the early

[70] Kreps, *U. S. Ocean Policy in the 1970s,* pp. ix-8.
[71] Personal communication, Office of Science and Technology policy staff personnel.
[72] Kreps, *U. S. Ocean Policy,* pp. ix-4.

1960s.[73] All of the major components of the ocean program have formed advisory committees, the advantages of which have appeared to outweigh the drawbacks over the years.

Traditionally, the National Academy of Sciences has been used as a handy source of scientific and technological opinion, especially by the federal agencies. While at first the only component of the academy relating to the ocean was the Committee on Oceanography, a series of reorganizations and formation of the National Academy of Engineering resulted in an expansion into the Ocean Sciences Board (the legacy of the original committee), the Ocean Policy Committee, the Marine Board, and the Maritime Transportation Research Board. These committees have conducted a number of useful reviews and analyses of various segments of the national ocean program. During the two-decade career of the ocean program, advisory functions have tended to blend with those of review; this may be helpful.

Generally speaking, however, a review of a given public organization or function serves as an advisory process to its supervisory body. Thus, in 1960, Dr. Wakelin, as chairman of the Interagency Committee on Oceanography, requested Dr. Athelstan Spilhaus, the chairman of the Committee on Oceanography of the National Academy of Science, to review the ocean plans and programs of the federal agencies and to advise him respecting the most and least likely activities to espouse.

Beginning in 1961, the President's Science Advisor each year invited a distinguished scientist to convene a committee of peers to review the ICO's program and to advise him accordingly. The PSAC panel fulfilled the same role in 1966 on a more intensive and comprehensive basis. Vice President Hubert Humphrey, as chairman of the Marine Sciences Council, convened review/advisory committees and panels to examine large numbers of subordinate organizations, programs, and issues, and finally assembled a master panel to assimilate and process the findings of the other panels.

The Stratton Commission (COMSER), established by law to advise the President and Congress, exceeded its predecessors in size, responsibility, and clout; its report is still referred to as the landmark analysis of the national ocean program.[74] Yet in 1974, Senator Ernest F. Hollings, as chairman of the Senate's Committee on Commerce, Subcommittee on Oceanography, succeeded in having the Senate charter a National Ocean Policy Study. Designed and conducted in the shadow of the Stratton Commission study, less than half a decade following dissolution of that distinguished body, the National Ocean Policy Study was apparently intended as a vehicle for rationalizing legislation of an independent ocean

---

[73] Price, *The Scientific Estate,* pp. 227-257.
[74] COMSER, *Our Nation and the Sea.*

department.[75] The Senator's proposal, the Department of Environment and Oceans Act, S 3889, died short of its apparent objective, failing to clear committee.

One of the most unusual of the ocean advisory groups was the original Sea Grant Advisory Panel, which constituted an equal mix of academia and industry at the level of president or senior vice president. The members were required to participate in two to four university site visits per year, in addition to their semi-annual business sessions. Overall, each contributed a full month of his year to the program. Due to the unusual and complex nature of the Sea Grant Program, the panelists considered that a minimum of two years apprenticeship was necessary to understand the program. It was probably the combination of individual competence and time devoted to the program which once led the NOAA Administrator, Dr. White, to describe the group as the toughest and most capable advisory body in the United States government.[76] Significantly, attendance at meetings appears to have averaged closed to 90 percent.

By the same reasoning, enactment of the Sea Grant Improvement Act of 1977 undoubtedly reduced the panel's effectiveness.[77] Political representation was partly substituted for talent as the criterion for panel membership; membership is currently limited to three years; attendance at panel meetings ranges from 25 to 50 percent, and the panel decisions and recommendations no longer dominate policy formation in the program.

The success of the Commission on Marine Science, Engineering, and Resources persuaded the Congress of the permanent value of such an instrumentality. Accordingly, in 1971, the Congress passed Public Law 94-69, creating the National Advisory Council on Oceans and Atmosphere (NACOA).[78] This council fulfills a top-level review function over all federal oceanic and atmospheric programs.

NACOA started out with an exceptionally experienced leadership: William Nierenberg, director of the Scripps Institution of Oceanography, chairman; William Hargis, director of the Virginia Institute of Marine Science, vice chairman; and Douglas Brooks, formerly president of the Travelers Research Center, executive director. The original group consisted of twenty-five world-class scientists, educators, and administrators, divided evenly among ocean scientists and atmosphere

[75] U. S. Congress, hearings before the Committee on Commerce; U. S. Senate, 91st Congress, Serial 91-59; March 23, April 2, 3, 14, 16, 21, 1970; pp. 820, 823, 850, 1007, 1009, 1025, 1046, and 1073.

[76] Personal communication from Dr. Robert White.

[77] Sea Grant Improvement Act of 1977.

[78] Public Law 94-69.

scientists. In later years, appointments appeared to depend more on political rather than scientific merit.

Each year the council submits a report to the President and Congress reviewing and offering recommendations concerning a few selected issues highlighting the National Ocean Program.[79] These analyses have included, among others: ocean engineering, outer continental shelf activities, fisheries, Sea Grant, coastal zone management, and the International Decade of Ocean Exploration. The council's *modus operandi* includes monthly meetings both in Washington and at other locations; internal deliberations and public hearings; and investigation by both full committee and subcommittee. Council membership has been recently reduced by the Congress from twenty-five to eighteen members, and delays in White House clearances have caused actual membership to drop to about a dozen.

The Situation Today

As the decade of the 1980s begins, the situation superficially appears serene. The National Ocean Program is larger, more expensive, and more complex than ever. The federal agencies with ocean programs appear to be making progress, held together at best as a loose confederation. However, if there is close communication among these agencies, it is not readily apparent. Leadership has been transferred in large part from scientists to lawyers, but this may be appropriate if, in fact, institutional problems have overtaken the technical ones.

The Navy, which has the longest tradition of supporting oceanography, appears to some to have relaxed its doctrine of relevance and adopted a slightly more liberal policy of supporting basic research not necessarily related to naval tactics and strategy.

The National Science Foundation appears to be maintaining leadership of the government's sponsored oceanographic research program. While the International Decade of Ocean Exploration (IDOE) will not totally end; in practice, the name will disappear, there will be a reduction of effort, and the funds hitherto assigned specifically for that program will be diffused throughout the other NSF ocean programs. IDOE served several functions. It demonstrated the value of big science in oceanography and that universities could coordinate under large-scale programs as pioneered by the Sea Grant Program. NSF and the National Science Board apparently take seriously their mandate to maintain an optimal mix of big and little ocean science in the national interest.

[79] Kreps, *U. S. Ocean Policy in the 1970s.*

Other than the Navy and the National Science Foundation, the number of agencies offering grants and contracts to universities has grown from a half a dozen twenty years ago to three times that number today. While this could be viewed superficially as accruing to the advantage of the grant seekers who now have many doors on which to knock, closer inspection reveals the fragility of this theory. The present duplication, confusion, and general disorder in the granting framework gives university administrators never-ending headaches. During the same period, the number of institutions educating oceanographers has grown from a couple of dozen to four times that number, and the employment market for some kinds of oceanographers must be classified as dismal.

The Environmental Protection Agency and the Corps of Engineers have become increasingly involved in coastal zone activity along with the National Oceanographic and Atmospheric Administration and myriad other federal agencies. The first major report linking these agencies in a master program of pollution research and monitoring has been completed and is now ready for production.

Overall, the Department of Commerce's major contribution to government planning has been the production of *U. S. Ocean Policy in the 1970s: Status and Issues.*[80] This was prepared in response to President Carter's request in June 1977 that the Secretary of Commerce conduct a comprehensive review of the United States ocean policy. It represented a large-scale assessment of the current ocean situation by the Office of the Secretary of Commerce.

During the past several years, the major related political activity has concerned federal reorganization and establishment of a Department of Natural Resources. In the face of mounting congressional resistance, the President has ceased his efforts in this direction. Accordingly, in the eyes of most political observers, the Washington ocean scene will be relatively quiescent for the next few years.

In the traditional sense of the word, coordination of the federal agencies' ocean effort is probably at the lowest level since the national oceanographic program began. There are several reasons for this. In the first place, the program itself has grown enormously in size and complexity. Second, as alluded to earlier, the National Oceanic and Atmospheric Administration, which was presumably established as the focus of federal leadership, actually has a smaller piece of the federal ocean pie than do several other agencies and, in fact, its proportionate aquatic charter appears to be diminishing with the years. By the same reasoning, the total funding, while increasing in the aggregate, becomes progressively diluted with the advent of new ocean and quasi-ocean

---

[80] Kreps, *U. S. Ocean Policy in the 1970s.*

programs, thus rendering conduct of the traditional programs more difficult over time. Third, there appears to be a change in emphasis from coordination in the "get together" sense to a focusing on issues. This appears to be the philosophy of the President's Science Advisor, Dr. Frank Press, and Dr. Richard Frank, administrator of NOAA and chairman of the existing coordination committee (the Committee on Atmosphere and Oceans). While cogent in concept, this type of administration obviously must depend for success on the most careful and adroit staffing. Yet, staff effort devoted to this function has, if anything, decreased to the point where the existing functionaries appear to be overloaded with work and overwhelmed with responsibility.

Accordingly, in terms of number of confrontations, joint address to problems and issues, joint analysis of programs and publication of reports, and joint execution and/or support of executed programs, coordination does, indeed, seem to be minimal.

There is formal trapping, however. At the time that the Carter administration received the first Congressional expression of the need for science policy coordination in the form of the act establishing the Federal Coordinating Council for Science and Technology Policy,[81] the White House attitude appeared (probably justifiably so) antagonistic toward councils, panels, and so forth. Accordingly, as previously mentioned, one of the council's early decisions on science coordination was to fuse subordinate coordinating committees on atmosphere and ocean, respectively, into a single function called the Committee on Atmosphere and Ocean (CAO). The chairman, appropriately, is Richard Frank, administrator of NOAA. An executive secretary serves without staff. At the executive office level, Dr. Phillip Smith, in the Office of Science and Technology Policy, has a general overview responsibility for ocean research and development, subordinate to Dr. Dennis Preger, executive secretary of the FCCST. Within the White House family, Dr. Catherine Schirmer of the domestic policy staff also exercises a degree of supervision of the ocean situation. By no technique of interpretation, however, do any of these persons appear to be considered either by responsibility or outlook, to be exhibiting the same paternal attitude toward the National Ocean Program that was typical of the Federal Council for Science and Technology, forerunner of two decades previous.

A fourth reason for the general malaise can be attributed to lack of accountability. The traditional Congressional discontent with the executive agencies' foot-dragging appears muted. This is attributed by some observers to the lack of new legislative heroes to replace the old guard, as they, in turn, either retired or went on to other things. At

---

[81] National Science and Technology Policy Organization and Priorities Act of 1976.

present, there does not seem to be any one person in either house who has adopted the mantle of leadership or assumed responsibility for rallying colleagues to the ocean cause. In the House of Representatives, John Murphy is chairman of the Merchant Marine and Fisheries Committee. Jerry Studds has assumed chairmanship of the Subcommittee on Oceanography, as John Breaux transferred to the Fish and Wildlife subcommittee. Other representatives display *ad hoc* interest in federal action that may affect their own constituencies. In the Senate, the attention of Senators Hollings and Magnuson appears to have been diverted to other interests, and long-time champions such as Senators Weikert and Pell, while still much interested, have become burdened with other major responsibilities. Accordingly, in the absence of the Congressional carrot-and-stick approach, there appears to be no sense of pressure on the Administration to highlight or tie in the ocean programs.

In another sense, the opinion might well be offered that Congress has, after all, done its job. The decade of the 70s has witnessed passage of at least a dozen landmark acts accelerating, encouraging, and generally beefing up the ocean program. How then to account for the paradox wherein the early 1960s appeared to bear the most vibrant and closely coordinated ocean movement in the executive branch of government, in the absence of the least material legislative action, whereas now, bolstered by strong and diverse statutes, the federal machine appears to be lumbering along without any particular sense of direction?

As usual, the answer is neither singular nor simple. First, having created the massive and complicated machinery, Congress lacks the analytical ability to assess it properly and to control it properly. Secondly, overwhelming national issues, such as environmental protection and energy engulf the ocean program like everything else, and subject it to pressures which are antagonistic to coordination. (Consider, for instance, the problem of meshing the ocean interests of the Environmental Protection Administration with those of the Department of Energy; yet, this is precisely what must be done).

Third, in a pragmatic sense, one of the products of Congressional action is constraint. In other words, the more detailed and comprehensive the legislative framework in which an executive agency operates, the greater the restrictions governing its activities, especially the extracurricular ones. Unfortunately, interagency coordination can be viewed as an extracurricular activity. Fourth, in a closely related view, the federal administrator whose enterprise is circumscribed by legislation will have an understandable tendency to look more often to the Congress for guidance, interpretation, and assistance, rather than to his fellow agencies.

Valedictory

A possibly unanticipated but significant result of the ocean program's growth in size, complexity, and awkwardness is the disappearance of focused responsibility. No longer can a chairman of the Interagency Committee on Oceanography, an ocean-minded Vice President, or even a NOAA administrator comprehensively answer the question: What is the state of our national ocean program? In fact, if there is a marine focus of national attention, it is more likely Jacques Cousteau, a Frenchman, than any of the invisible but politically potent individuals or agencies in Washington, D.C.

Finally, it is difficult to argue these or any other statements in the absence of a public report on the status, progress, and or accomplishments of the national ocean program. On faith, one must necessarily suppose that it is alive and (fairly) well, and living somewhere in Washington, D.C.

ADVOCACY

CHAPTER 2
THE MAJOR UNITED STATES
FEDERAL GOVERNMENT MARINE
ORGANIZATION PROPOSALS

ROBERT E. BOWEN*

Introduction

Recent technological advances, the recognition of potential terrestrial resource shortages, and a generally expanded awareness of our ability to impact the marine ecosystem have begun to push our sphere of political interest seaward. Oil and gas reserves are becoming increasingly important to world energy supplies. The potential billion dollar contribution of manganese nodules to international commodity markets has heightened our interest in the deep seabed. The need to manage the utilization of fisheries stocks to prevent overexploitation is becoming more widely recognized. We are, in short, much more conscious of the sea as a political and economic entity than we were a quarter of a century ago. Accordingly, effective management of marine resources has become a more central concern for national governments. Jurisdictional extensions over economic activity in nearshore waters is one manifestation of this awareness. The extended debate in the Third United Nations Conference on the Law of the Sea can be viewed as an attempt by the international community to define an equitable and efficient scheme for managing global marine resources.

However, in most cases this shift in national perception has not been matched with an attendant change in the bureaucratic structures involved in the design and execution of marine policy. This is at least partially true in the United States. Although numerous organizational proposals have emerged during the past dozen or so years, little action has been taken by the government.

In this chapter I will review the major reorganization proposals as well as present an overview of the current federal organization structure for the making of United States ocean policy.

* Robert E. Bowen is a research assistant, Institute for Marine and Coastal Studies, University of Southern California.

The Status Quo

The idea of restructuring the federal bureaucracy is not unique to marine affairs. Indeed, some argue that the concept of reorganizing the executive branch is one of the few constants in the history of our government. The five original departments created in 1789 have since grown to 12 cabinet-level departments; 58 agencies, commissions, independent regulatory agencies, administrations, authorities, corporations, boards or services; and three quasi-federal agencies. Ocean programs are administered by approximately two thirds of those administrative units.[1]

Figure 1[2] is a diagram of the organization of departments and subagencies presently responsible for marine programs and services. From this chart, it is apparent that the bureaucratic range in which marine affairs are administered is remarkably broad. However, the central focus here lies with the National Oceanic and Atmospheric Administration in the Department of Commerce. NOAA was created in 1970 to "exercise leadership in developing a national oceanic and atmospheric program of research and development."[3] Since 1970, NOAA has assumed a number of additional responsibilities, primarily regulatory and developmental in nature; thus, the mission of NOAA has slowly changed. NOAA is generally perceived to be fairly successful in carrying out these added duties. However, as a coordinating force it is viewed as being less successful. For instance, between 1969, the year before NOAA was created, and 1977 the number of administrative departments in which ocean programs were located nearly doubled. This trend has led to what many feel is an undue amount of jurisdictional overlap and duplication of programs. Responsibility is shared by two or more agencies in most major ocean activities and roles assigned to federal government.[4] Three interesting cases serve as examples of this. For instance, although regulation of marine fisheries falls within the jurisdiction of NOAA in the Department of Commerce, the management of anadromous species is carried out by three cabinet-level departments. Because these stocks spend part of their life cycle in fresh water and part in saltwater, they are regulated by NOAA, the Fish and Wildlife Service of the Department of the Interior, and the Department of State.

---

[1] U.S. Department of Commerce, *U.S. Ocean Policy in the 1970s: Status and Issues*, October, 1978, pp. ix-1.

[2] *Ibid.*, pp. ix-36.

[3] U.S. Congress, House Committee on Government Operations. *Reorganization Plan No. 4*. Hearings before a subcommittee of the House Committee on Government Operations. 91st Cong., 2nd Sess. 1970, p. 8.

[4] U.S. Ocean Policy, pp. ix-22.

Oil and gas platforms in coastal waters with pipelines running ashore could involve regulation by the Department of Energy, the Interstate Commerce Commission, the Department of Transportation, the U.S. Army Corps of Engineers, and the Environmental Protection Agency.

Perhaps the most interesting example of this kind of division of responsibility is the case of the green sea turtle. Through most of the life span, the turtle swims along the coast and in the open sea. During this time, it is under the jurisdiction of NOAA. However, once every two to four years, the female turtles move onto the beach to lay eggs. During those periods, and when they are occasionally found basking in the sun along the coast, green sea turtles are the responsibility of the Fish and Wildlife service of the Department of Interior.

Although it can be argued that these problems of overlapping jurisdiction difficulties are currently more widespread, many authors note that they have been in existence for some time. This has been at least partially the source of a rather steady stream of alternative organizational schemes from members of the ocean community. It is beyond the scope of this study to review all of these proposals. Instead, I will discuss six of the more prominent and innovative schemes. The first four are primarily historical documents: the Stratton Commission report, the Ash Council report, the Moore Proposal of 1976, and the Hollings Proposal of 1976. These four schemes collectively exemplify much of the conventional wisdom in organizing the government for the effective execution of its maritime policy. The section will conclude with detailed discussion of the two major proposals that are currently being considered: (1) The Carter Reorganization Project/Department of Natural Resources recommendation of 1978; and (2) the 1979 proposal of the National Advisory Committee on Oceans and Atmosphere.

The Stratton Commission (1969)

Undoubtedly the best known effort for reorganization was that proposed by the Stratton Commission. Formally, known as the Commission on Marine Science, Engineering and Resources, the commission was empowered by Congress to conduct a "comprehensive investigation and study of all aspects of marine science in order to recommend an overall plan for an adequate national oceanographic program that will meet the present and future national needs."[5] Under the able chairmanship of Julius A. Stratton, president emeritus of the Massachusetts Institute of Technology, the commission produced its final

[5] P.L. 89-454, Sec. 5(b). Appendix 7.

Figure 1

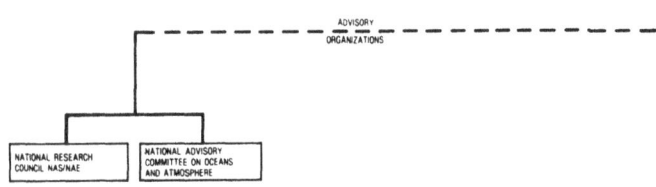

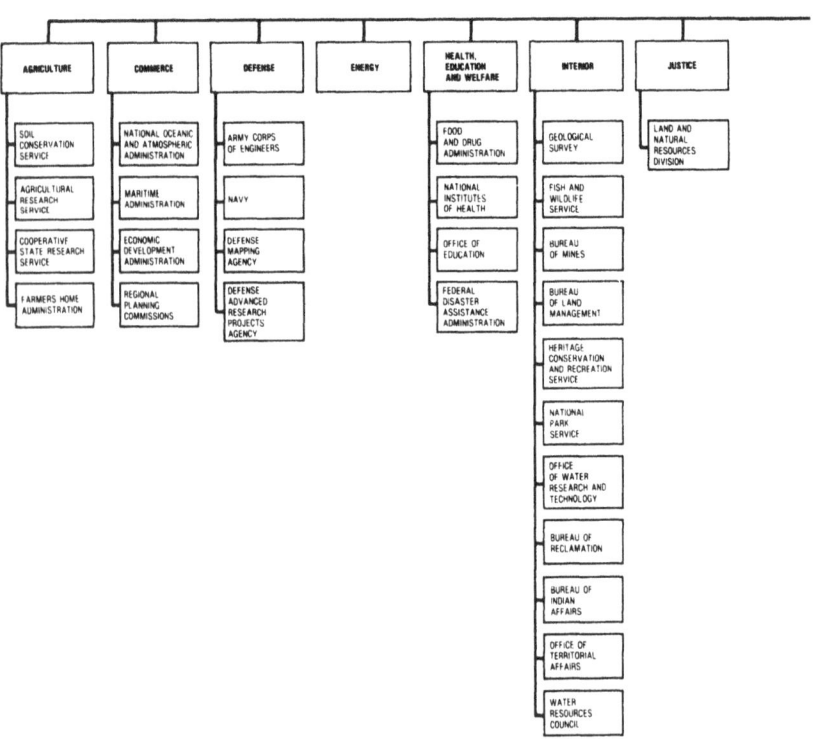

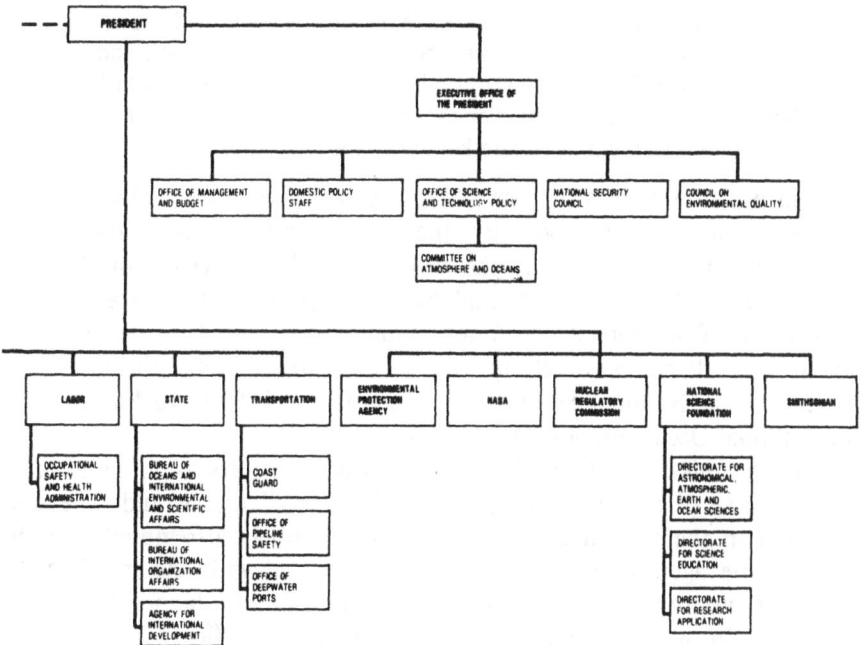

report in 1969. Entitled *Our Nation and the Sea*[6] it stressed the growing need for the country to more rationally manage the wide range of marine resources and made a series of policy recommendations. At the core of these recommendations was the call for the creation of a new independent agency to coordinate marine-related activities:

> It is our conviction that the objective of the national ocean program recommended by this commission can be achieved only by creating a strong civil agency within the Federal Government with adequate authority and adequate resources. No such agency now exists, and no existing single Federal agency provides an adequate base on which to build such an organization. For the national ocean effort, we propose unified management of certain key functions is essential.

Figure 2 shows the structure of the National Oceanic and Atmospheric Agency that was proposed by the Commission.[7] The Commission recommended essentially that the new agency be composed of the U.S. Coast Guard, the Environmental Science Service Administration, the Bureau of Commercial Fisheries (augmented by the marine and anadromous fisheries functions of the Bureau of Sport Fisheries and Wildlife), the National Sea Grant Program, the U.S. Lake Survey, and the National Oceanographic Data Center.

The Commission felt that NOAA, constituted as proposed, could bring a "freshness of outlook and freedom of activity" that would otherwise not exist in the federal bureaucracy.[8] The members also noted that the head of the agency would be better able to deal with the multiple-use questions inherent in marine affairs, and would be in a favorable position to aid the President in coordinating the federal ocean effort.

The document was widely circulated and was generally quite favorably received by the ocean community. Various difficulties, primarily bureaucratic, prevented early acceptance of the recommendations. The call for a strong independent NOAA fell under fire from several cabinet-level departments. However, perseverance on the part of supporters of the report did, in the end, result in at least partial acceptance by the administration. In July of 1970, President Nixon transmitted to Congress Reorganization Plan No. 4,[9] in which the creation of a National Oceanic and Atmospheric Administration was proposed. The President's proposal followed the recommendations of the Stratton Commission except in three key areas. First, NOAA was placed within the Commerce

---

[6] Commission on Marine Science, Engineering, and Resources. *Our Nation and the Sea: A Plan for National Action* (Washington, D.C.: Government Printing Office, 1969), p. 229.

[7] *Ibid.*, p. 233.

[8] *Ibid.*, p. 233.

[9] Reorganization Proposal, No. 4; Appendix 10 of this proposal.

Figure 2

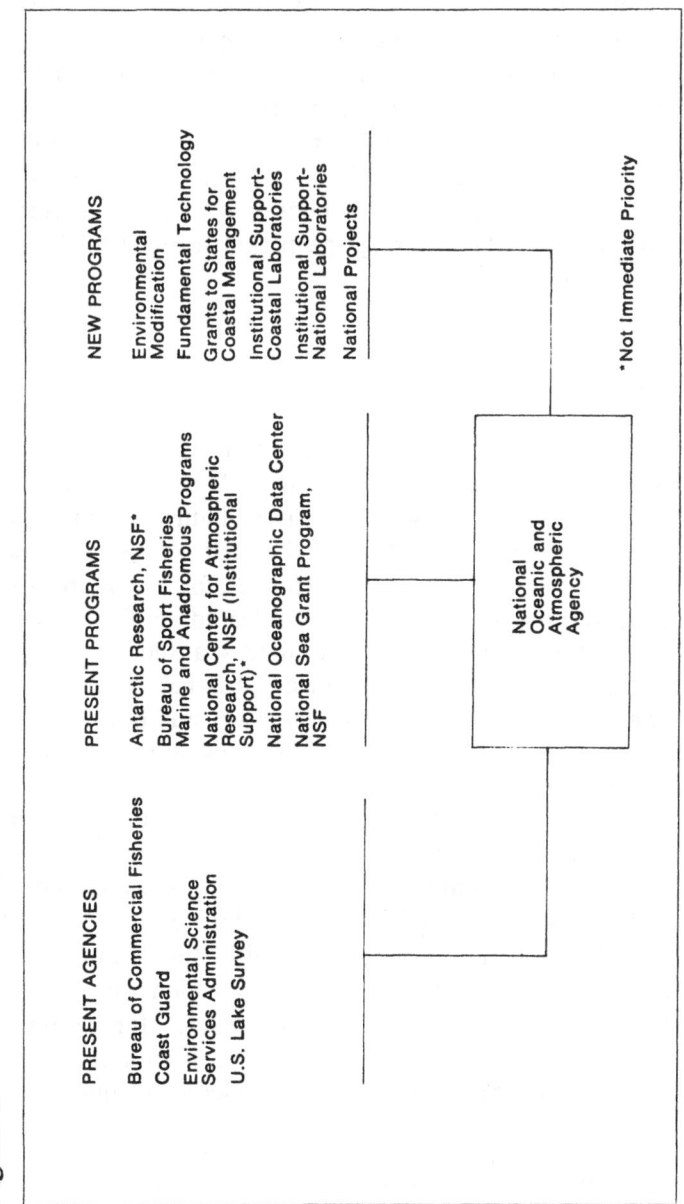

PRESENT AGENCIES

Bureau of Commercial Fisheries

Coast Guard

Environmental Science
Services Administration

U.S. Lake Survey

PRESENT PROGRAMS

Antarctic Research, NSF*

Bureau of Sport Fisheries
Marine and Anadromous Programs

National Center for Atmospheric
Research, NSF (Institutional
Support)*

National Oceanographic Data Center

National Sea Grant Program,
NSF

NEW PROGRAMS

Environmental
Modification

Fundamental Technology

Grants to States for
Coastal Management

Institutional Support-
Coastal Laboratories

Institutional Support-
National Laboratories

National Projects

National
Oceanic and
Atmospheric
Agency

*Not Immediate Priority

Department, thereby ignoring the commission's call for independent status. Second, it did not incorporate the Coast Guard which remained in the Department of Transportation, and third, the new functions proposed in the 1969 report, particularly those centering around marine technology, were not granted to the new ocean agency. It was hoped that this "unified approach to the problems of the oceans and atmosphere" and the creation of NOAA would force a coordination of effort in the national ocean effort. However, as noted earlier, strong central coordination of marine-related projects within the federal government is far from a reality. Index estimates produced by the General Accounting Office (GAO) indicate that only 7.9 percent of the total budget related to marine science activities and oceanic affairs during the period from 1970 through 1974 was budgeted to NOAA.[10]

The Ash Council (1971)

Early in his first term of office, President Nixon created the President's Advisory Council on Executive Reorganization. Under the leadership of Roy L. Ash, president of Litton Industries, the council was designated to examine the structure of the staff offices surrounding the President and to evaluate alternative schemes for the reorganization of the federal government. Its report was released in February 1971. In March, the President submitted to Congress a federal reorganization program that attempted to institutionalize a number of the Ash Council recommendations.[11] The reorganization program was broadly defined and called for eight cabinet-level agencies including a new Department of Natural Resources. One of the administrative units within the DNR was to be an Oceanic, Atmospheric and Earth Science Administration, which was to be formed by the joining of NOAA and the U.S. Geological Survey. Little specific attention was given to the oceans, and the basic organizing theme was the need to end the duplication of scientific data gathering and dissemination services provided by the two organizations.

The 92nd Congress took no legislative action on the proposed Department of Natural Resources. However, the proposal was reconsidered by President Nixon and was resubmitted to the 93rd Congress in a slightly modified form.[12] In reaction to questions on the

[10] U.S. Ocean Policy, pp. ix-1.
[11] U.S. Office of Management and Budget. *Papers Relating to the President's Departmental Reorganization Program: A Reference Compilation* (Washington, D.C.: Government Printing Office, 1971), p. 172.

adequacy of the nation's energy resources, the new program called for the creation of a Department of Energy and Natural Resources (DENR). There was little difference between the 1971 proposal for a Department of National Resources and the 1972 proposal for a Department of Energy and Natural Resources. Once again, no legislative action was taken.

The Moore Proposal (1976)

John Norton Moore, Director of the Center for Oceans Law and Policy at the University of Virginia, has also offered an interesting and in many ways unique proposal. Prepared in 1976, his paper stressed the need to "ensure an enduring organizational structure capable of developing and refining ocean goals and priorities and of effectively managing oceans programs."[13] There were three major components to his plan. First, he proposed the creation of a cabinet-level Marine Affairs Council, which would be chaired by the Vice President of the United States and which would develop national ocean goals and would, on behalf of the President, oversee ocean programs. The council would act as a key force in centralizing and coordinating the development of national marine goals. The proposed council was quite similar in structure to the National Council on Marine Resources and Engineering Development, known as the Marine Science Council, which was formed in 1966 at the same time as the Stratton Commission. Many analysts have noted that with Vice President Humphrey as chairman, the Marine Science Council was at least relatively effective in exercising leadership in the development of marine policy.

The second component of the Moore proposal dealt with the restructuring of NOAA. He suggested that NOAA be moved out of the Commerce Department and be granted independent status. It should include:

> . . .the Coast Guard, the Maritime Administration, the Outer Continental Shelf and deep seabed mining programs of the Interior Department, the Marine and Coastal Zone activities of the Army Corps of Engineers, most oceans research Programs of the National Science Foundation (including the International Decade of Ocean Exploration), most oceans research and monitoring Programs of the Environmental Protection Agency, the oceans and atmosphere

[12] U.S. Congress, Senate, Committee on Interior and Insular Affairs. *Congress and the Nation's Environment.* 94th Congress, 1st Sess., 1975, p. 746.

[13] John Norton Moore, "Organizing for a National Oceans Program," in *Oceans 76, Proceedings of the Marine Technology Society,* September 13-15, 1976, 1E-2.

activities of the Bureau of Reclamation, and some ocean related activities of the Fish and Wildlife Services, particularly programs for anadramous species and marine mammals.[14]

In structuring NOAA in this fashion, and in particular in transferring the Maritime Administration to an expanded and independent agency, Moore went beyond even the Stratton Commission report in recommending centralization of the national ocean effort within a single administrative unit.

The third recommendation made by Moore was to reorganize and strengthen the state department's ocean effort by creating a Bureau of Oceans and Environment. Thus by centralizing the state department's effort, Moore felt that greater levels of efficiency in defining and carrying out our marine policies in the international arena were possible. No legislative action was taken on any of the recommendations from Moore's proposal.

The Hollings Proposal (1976)

The next proposal to be reviewed here is that of Senator Ernest Hollings. As chairman of the Senate National Ocean Policy Study, Senator Hollings oversaw several years of discussions on ocean and environmental affairs. One product of these debates was the introduction, in October 1976, of S3889; a bill that called for the creation of a new Department of the Environment and the Oceans (DEO). Into this cabinet-level department would be transferred the Environmental Protection Agency (EPA), NOAA, the Coast Guard, and a number of services and programs from both the U.S. Army Corps of Engineers and the Department of the Interior.[15]

In his comments at the time the bill was introduced, Senator Hollings identified the organizing principles around which the bill was drafted. Indeed, it was the framework of the proposal that set it apart from other administration and congressional efforts. Most important in this regard was the recognized need for the federal government to deal comprehensively with what are called the "Nation's common property resources."[16] Common property resources are unique conceptually in that once they are produced it is not possible to preclude others from sharing

[14] Ibid., pp. 1E-2.
[15] U.S. Congress, Senate, Committee on Commerce, Science and Transportation and National Ocean Policy Study. Congress and the Oceans: Marine Affairs in the 94th Congress. 95th Congress, 1st Sess., 1977, p. 308.
[16] 122 Congressional Record S17854 (daily edition, October 1, 1976).

in their consumption. Questions of environmental protection and marine resources quite clearly fall within this conceptual framework. Once again, Congress took no action on this.

The President's Reorganization Project/Department of Natural Resources (1978)

One of the most recent major schemes to reorganize the national ocean and atmospheric effort was President Carter's plan for a new Department of Natural Resources (DNR). A central theme in Carter's 1976 presidential campaign was the need to reduce waste and inefficiency in the federal government. Toward these ends, he created the President's Reorganization Project (PRP) to review the state of the federal bureaucracy, to specify problem areas, and to identify appropriate solutions. The DNR was one of a number of governmental reorganization schemes that emerged from PRP.

The general purpose of the new cabinet-level department was to bring under a single administrative jurisdiction the major natural resources development problems faced by the nation. That concept is not particularly novel; indeed, similar proposals have appeared periodically through a number of administrations. The version utilized for this study is the PRP staff analysis, Natural Resources Reorganization, dated December 16, 1978:

Proposal

Form a Department of Natural Resources with the mission of: managing the nation's natural resources for multiple purposes, including protection, preservation, and wise use.

Composition of the New Department

All of the Department of Interior except the construction functions of the Bureau of Reclamation (which would go to the Army Corps of Engineers, together with the construction functions of the Soil Conservation Service's water shed program); the National Oceanic and Atmospheric Administration from the Department of Commerce; the U.S. Forest Service, and the water shed planning and soil and snow surveys of the Soil Conservation Service from the Department of Agriculture; the preauthorization and preconstruction planning and budgeting from the Corps Engineers civil works; and the Water Resources Council.

Cost Savings

Substantial cost and personnel savings would be possible through the merger of similar functions, streamlining of internal organization of the Department, efficiencies in service delivery, and the elimination of duplicative activities.[17]

The staff analysis includes a number of statements that are indicative of general background and reasoning:

The proper federal role is one of balancing competing claims for the use of public resources (energy, food, recreation, transportation, timber, minerals, and scenic or historic values) as well as balancing use versus often conflicting values of preservation and quality protection. The federal resource management role also requires careful balancing of present-day pressures with future needs in order to anticipate and forestall crises and assure that future needs can be met. The need for balancing involves collaboration with the States in reconciling national needs. Finally, it requires a knowledge of the citizens' perspectives at regional and local levels who are strongly affected by decisions about resources in their area. The federal institutions responsible for fulfilling this role were designed one by one in a previous era to address specific problems associated with the settlement of the West and the production of raw material for an expanding economy. New organizations were established on a piecemeal basis for narrow purposes such as encouraging settlement of areas of the public domain, providing irrigation water to farms in arid areas, managing and conserving timber, or locating and developing energy and mineral resources. Over time, as the nation's needs changed, the scattered institutions have been given broader responsibilities by Congress. This has evolved into a system that lacks comprehensiveness and, at the same time, contains overlaps. It produces delays, confusion and inconsistencies, excessive costs, and narrowly based decisions. No one in the present system is accountable for addressing natural resources issues in a comprehensive way even though extensive interactions within the physical world are generally recognized and need to be addressed.[18]

Present-day problems which arise from the adequate structure that has evolved historically, will intensify in the future. Resource stewardship issues and decisions will increase in numbers and importance as the pressures on our limited and fragile resources grow. Increasing population, economic growth, rising living standards and increasing demand for outdoor recreation will further frustrate the situation.

---

[17] Office of Management and Budget. President's Reorganization Project. *Staff Analysis: Natural Resources Reorganization.* December 16, 1978, p. i.

[18] *Ibid.,* pp.1-2.

The DNR would bring together in one cabinet-level department the necessary components for balanced natural resources management.[19]

The proposed DNR structure would be based on five operating administrators (or equivalents), each responsible for executing a discrete but interrelated portion of the total responsibilities assigned to the Secretary. Secretarial offices would have cross cutting and/or staff responsibilities and would work closely with the operating components to aid the Secretary in carrying out his policies and programs.[20]

The five proposed Administrations are as follows: Oceanic and Atmospheric Resources; Land Resources; Water Resources; Resource Sciences and Data Monitoring; Parks, Recreation, Heritage and Wildlife. The proposal assumes that federal responsibilities for Indian Affairs and territories would continue to be closely related to the federal natural resources functions and would thus be included in the DNR.[21]

Figure 3 shows the organization advocated in the proposal.[22] The proposal then turns from the general structure of the DNR to specific analysis of the roles of the oceans and atmosphere in the overall scheme. In particular, the staff memo addresses the problem areas that DNR would overcome.

Ocean and Atmospheric Resources

The rapidly expanding federal role in managing ocean and atmospheric resources is split between the National Oceanic and Atmospheric Administration (NOAA) in Commerce and Interior. NOAA is predominantly a research and scientific agency and is striving to implement new management and regulatory authorities. Interior has the authority for managing the mineral and energy resources of the Outer Continental Shelf (OCS) and expertise in marine mammals and anadromous fisheries. Other federal agencies have lesser roles.[23]

1. There is no effective focal point in the federal government for developing ocean resource policies or for providing comprehensive management of ocean-related natural resources. As a result, the process of developing coherent oceans resource policies that reflect the broader natural resource view of the nation has been difficult to achieve.

[19] *Ibid.*, p. 4.
[20] *Ibid.*, p. 4.
[21] *Ibid.*, p. 5.
[22] Office of Management and Budget, President's Reorganization Project, *Proposal for Natural Resources Reorganization*, April 6, 1979, p. 3.
[23] Staff Analysis: Natural Resources Reorganization, pp. 9-10.

64

Figure 3

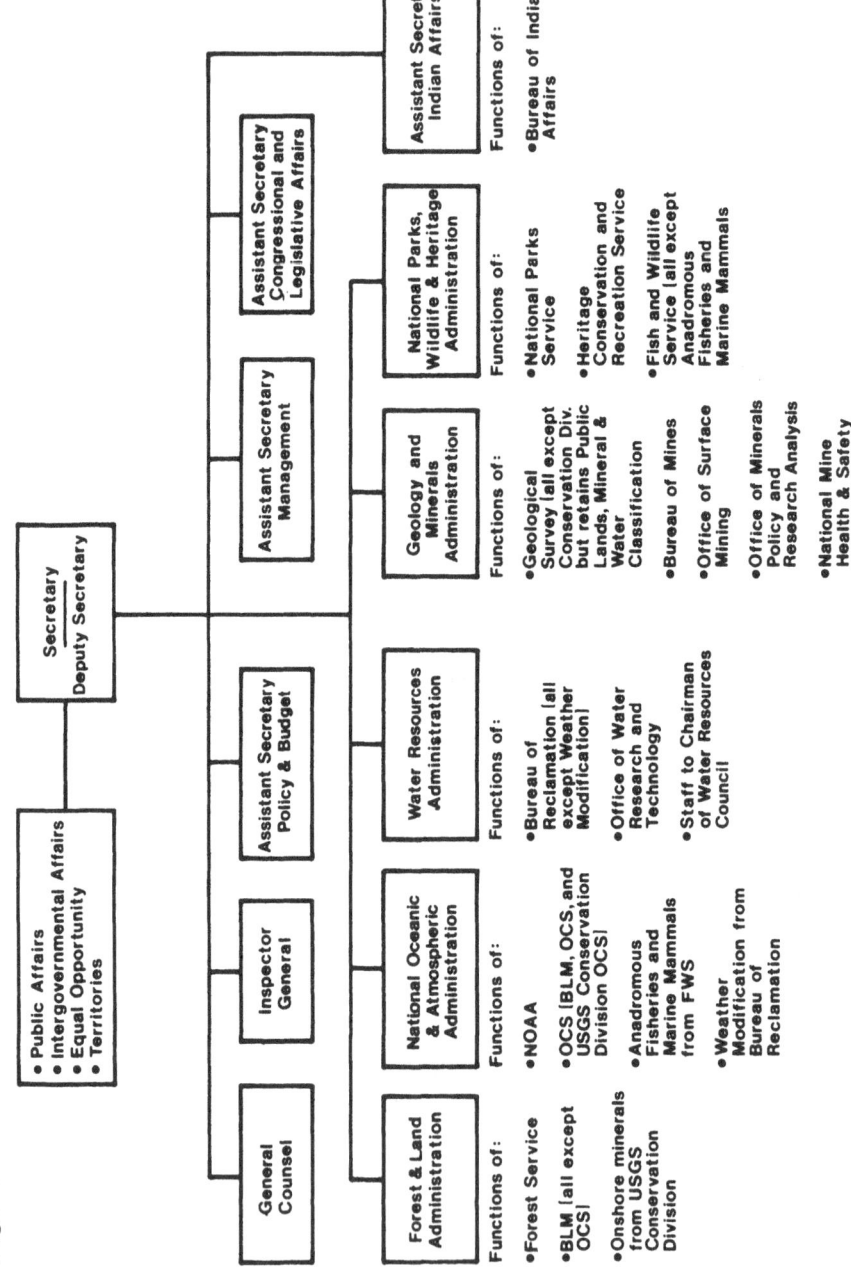

2. The split responsibility among Interior, NOAA, the Department of Energy and EPA for development and management of OCS oil and gas resources and for the protection of the marine environment has encouraged delays and legal battles, and led to increased expenditures for environment baseline studies.

3. An administration position on the appropriate lead agency for deep seabed mining remains unsettled. Interior and Commerce are seeking the lead role—the rule is wasted effort, unproductive "turf" fights, and a lack of accountability.

4. Ocean resource development, including OCS oil and gas, fisheries, deep-water ports, aquaculture and recreation, all impact the coastal zones. Federal responsibility for development of these resources is dispersed among several agencies, making required federal consistency with state coastal zone management plans more difficult to achieve.

5. Split reponsibility between Interior and NOAA has hindered implementation of endangered species and marine mammal protection legislation, resulting in delays in decision making. For example, promulgation of regulations governing the sea turtle was delayed for three years by jurisdictional disputes.

6. Weather modification research is found in several agencies including NOAA and Interior. Results thus far have been rather meager due to the dispersion of resources and accountability. This mismanaged area of Science, Technology, and Applications has been described as "fragmentation" in ten or more studies by a variety of groups over the past 12 years.[24]

According to the report, these difficulties can be overcome by the following program changes:

Transfer NOAA to the DNR and use it as a basis for creating an oceans and atmospheric agency within the DNR. This new agency would have the responsibility for oceans and atmospheric policy making, OCS management, deep seabed mining, coastal zone management (CZM), weather programs, marine geology, and physical oceanography. This action would also consolidate in one department (the DNR) existing NOAA and Interior responsibilities related to marine mammal protection, endangered species, and anadromous species programs.[25]

---

[24] *Ibid.*, p. 10.
[25] *Ibid.*, p. 10.

National Advisory Committee on Oceans and Atmosphere (1979)

The National Advisory Committee on Oceans and Atmosphere (NACOA) was created in 1971 to provide advice and to serve as a comprehensive oversight committee to Congress and the President on marine and atmospheric affairs. It is probably the one body that has been most active and consistent in addressing the problems of organizing the national ocean effort. Indeed, NACOA has expressed itself on the need for examining the organizational difficulties of the federal program in each of its seven annual reports. During 1978, NACOA sponsored two workshops designed to specifically address the question of federal reorganization of ocean affairs. The first workshop was limited to representatives of federal agencies while the second included participants from the academic, professional, and business communities. The discussions and debates in those workshops culminated in a report of findings and recommendations which offered the following:

> NACOA finds that the United States Government is not now organized to carry out its domestic and global responsibilities adequately if this nation and the world are to: 1) achieve the full economic potential of the users of the sea; 2) manage the global aspects of environmental protection; and 3) acquire and apply the scientific understanding and advanced technology needed for these purposes and for exploring, predicting, and, where feasible, controlling the complex phenomena of the oceans, atmosphere, coastal zone, and polar regions.
>
> NACOA finds also that the higher technology base of NASA and the global nature of its activities offers an intriguing potential for solving increasingly critical problems by the combination of NASA activities with those of the oceans and atmosphere.
>
> NACOA recommends that the President and the Congress undertake to reorganize the non-military federal structure dealing with the oceans, atmosphere, coastal zone, and polar regions so as to centralize the management of its activities relating to the economic development of the uses of the ocean and associated coastal areas, the protection of the marine and coastal environments, and science and technology related thereto.
>
> NACOA recommends specifically that the President and the Congress establish a new department responsible for atmospheric, oceanic, coastal, and polar affairs that would incorporate the bulk of the missions and programs presently associated with these activities. The department would include the National Oceanic and Atmospheric Administration (NOAA), and the United States Coast Guard from the Department of Transportation. Also included would be programs

involving marine fisheries and aquaculture, marine mammal protection, the preservation of endangered marine and coastal species and habitats, offshore oil and gas development, ocean mining, and some aspects of coastal engineering and of regulation for the protection of marine and coastal environment, as well as programs on climate research and weather modification. These programs are now split among the Departments of Commerce, Interior, Agriculture, and Transportation and the Army Corps of Engineers. Not included are programs directly related to the basic missions of the Department of State, the Department of Defense, the Environmental Protection Agency (EPA), the Council on Environmental Quality (CEQ), and the National Science Foundation (NSF).

NACOA recommends also that the President and the Congress evaluate the desirability of including all or part of NASA's activities in the new department.[26]

In support of their position, NACOA expressed concern that present state or federal organization is not adequate to meet present national needs with regard to the oceans and atmosphere.

Globally, we need to help build new institutions governing seagoing operations that are conducive to international harmony and cooperation. We must seek a global future of shared rather than conflicting interests, but one that will at the same time serve our domestic self-interest. Domestically, we need to reverse or contain the adverse trends which limit the potentially large economic returns from our ocean activities, threaten many living resources, jeopardize the integrity of the ocean environment, and produce an increasing gap in what we know about the planetary envelope of ocean and atmosphere and what we need to know to carry out our objectives. Our obligation is at the same time to ourselves and to the world. . . .[27]

This is the global challenge that must be met. Yet meeting this challenge requires more than a collection of programs each dedicated to some phase of the problem. It requires the formulation of an overall policy dedicated to progress on a broad front and an organization to implement such policy effectively and efficiently. The current consideration of government by the Administration is timely and provides an opportunity to consider how best to bring the potential of the oceans to fruition. Accordingly, NACOA, as is its statutory duty, has undertaken to formulate measures for meeting this global challenge of the oceans as a national imperative.[28]

[26] National Advisory Committee on Oceans and Atmosphere. A Special Report to the President and the Congress. *Reorganizing the Federal Effort in Oceanic and Atmospheric Affairs*. February 1979, p. 1.

[27] *Ibid.*, p. 3.

[28] *Ibid.*, p. 5.

In specifically addressing the shortcomings of the present structure, the NACOA report identified a number of areas in which the problems had been most acutely felt.

1. Lack of an integrated ocean policy—a virtual impossibility with the growing oceans program fragmentation and the absence of a high-level Executive Branch focus for oceanic matters;

2. Inability to resolve promptly such important conflicts as that between outer continental shelf (OCS) oil development interests and environmental protection advocates in order to avoid the inordinate and costly delays equally frustrating to all;

3. Inadequate conflict resolution policies and mechanisms to strike a proper balance among competing and incompatible uses of the crowded regions offshore;

4. Lack of an aggressive pro-economic ocean use development policy and program focus;

5. Absence of a sufficiently integrated policy for environmental protection and resource conservation and a focused, well informed regulatory activity to implement it; and

6. Insufficient support for a strong science and technology arm required for the development, management, and regulatory programs of the Federal Government aimed at fully effective use of the oceans, the atmosphere, and the coastal and polar regions.[29]

In stressing the systematic view of marine and atmospheric issues that are central to NACOA's analysis, the report concludes that:

A global outlook is imperative for overseeing the national interest in the oceans. This insular view, expertise, and programs needed for management of land and fresh water resources are vastly different from those needed to deal with conditions at sea. It is, of course, necessary that oil and gas development on the continental shelf, like that on land, be tested in accordance with our overall national energy policy and plans. However, the conditions of development must reckon with the potential cross impacts on shipping, fish and fishing, national defense installations, and the beaches and wildlife sanctuaries. All this, as well as the design, layout, and location of shoreside developments induced by ocean activities, requires special engineering, environmental, operational, legal, economic, and international policy analysis, expertise, and cooperation. This is true

[29] *Ibid.*, p. 8.

whether we have commercial, environmental, or scientific goals in mind.[30]

One section in the NACOA report details the reaction of the committee to the President's DNR proposal.

> We question whether a Department of Natural Resources (DNR), as is being considered, can manage this range of U.S. interests and at the same time deal adequately with its new responsibilities for land-based natural resource management, which will in themselves be difficult, complex, and demanding. We fear that a DNR Secretary would be unable to give adequate attention to both land and sea, particularly as he attempts to bring order out of the inevitable strains of reorganization. We believe, of the two, ocean interests are in greater need of special attention because the conditions in which they must cope are changing faster than those facing land resource development, and they are starting from farther behind. In the final analysis the proposition of integrated ocean resource programs within a Department of Natural Resources, though superficially attractive, is inadequate for realization of the full national interest in the potential of the oceans.[31]

Final Comments

This short descriptive essay has reviewed six of the major proposals that attempted to design what the authors felt would be a more effective federal structure for the definition and execution of the nation's marine policy. While there are some rather basic differences between the proposals, a number of themes are common to all six studies. For instance, there is certainly a common general perception of the inadequacy of the federal organization. Jurisdictional overlap and duplication of programs are problems noted in all of the proposals. The need for an administrative focal point in the identification of marine policy goals is also common concern. Furthermore, most of the proposals stress the need for a more viable ocean advocate with greater direct access to the President.

Despite the degree of commonality in the perception of the problem, the solutions vary widely. Nevertheless each of the proposals adheres to one of three organizing concepts. In the first concept, the oceans are viewed as the central organizing principle. The problems of managing marine resources are perceived as being sufficiently unique and important

---

[30] *Ibid.*, p. 8.
[31] *Ibid.*, p. 40.

to warrant special consideration. These concerns have been manifest in the call for a strong independent ocean agency or cabinet-level department. The Stratton Commission Report, the Moore proposal of 1976, and the 1979 NACOA proposal are examples. The creation of NOAA in 1970 can be viewed as partially legitimizing this view.

In the second concept, the oceans are viewed as being subordinate to the larger question of natural resource management. One can view the Nixon and Carter proposals calling for the creation of a Department of Natural Resources in this context. In both of these schemes, ocean issues received relatively little emphasis, and questions of marine policy would be primarily addressed within the decision-making structure of a larger cabinet-level department.

The utilization of common property resources as a conceptual focus is a third theme. While the Hollings Proposal is the best example, this theme has run at least subliminally through much of the reorganization debate.

Given the number and range of proposals that have emerged, why has the present federal structure remained relatively unchanged. A number of factors contribute to this situation. Among them are a rather wide range of organizational situations, cumulatively termed "organizational inertia," that make it quite difficult to seriously consider the oceans as a separate focus. Additionally, a degree of ambiguity surrounds the economic exploitability of certain marine resources, particularly deep seabed manganese nodules. This appears to have introduced sufficient uncertainty to at least postpone any action.

Actually, however, the answer is basically simple. The oceans are not generally viewed as being important enough to serve as the focus for a major effort toward federal reorganization. As the sea and its resources are perceived as being more central to the nation's vital interests, government perceptions will likewise change, and it will organize and act to account for these changes. Until then, it is likely that the oceans will not act as a major force in federal reorganization.

CHAPTER 3
NATIONAL ORGANIZATION
FOR OCEAN MANAGEMENT:
CENTRALIZATION VS. FUNCTIONALIZATION

DON WALSH*

Introduction

In the United States, the whole area of ocean policy and management has become very trendy in the past five or six years. Almost every federal agency with any ocean activities, as well as both houses of Congress, have in progress some sort of ocean policy and management study. However, the fundamental paradox is simply this: you cannot be objective if you are one of the affected parties. It is hard for a doctor to give himself a physical examination.

This is not to say that the multiple study efforts now being done in federal agencies and in Congress are not useful. These are important in helping each constituent element understand its possible future role in a new national ocean program initiative. Also, a valuable corollary purpose is served in training agency management to think about ocean issues as a part of their organization's future direction. This overall sensitizing of an increasing number of policy people in the executive and legislative branches of government should help in future development of a national ocean policy and management structure. However, there clearly needs to be supra-agency integrating study efforts which balance out the natural but self-promoting aspects of the in-house studies. We must look to agencies outside government for this most vital perspective. For example, these might be the National Research Council, universities, and non-government research organizations. The basic issue here is *sea power*. Admittedly, this term is badly misused, especially since the end of World War II. But in the context that was intended by the great nineteenth century maritime strategist, Admiral Alfred Thayer Mahan, it is quite applicable to the current situation. For this chapter, I will define sea power as "The sum total of national uses of the sea." That is, how do we effectively use ocean space as an instrument of our national policies?

At the end of World War II, the United States emerged as the greatest sea power the world had known. It had the largest navy; the greatest

* Don Walsh is Professor of Engineering and Director of the Institute for Marine and Coastal Studies at the University of Southern California.

number of merchant ships; the second largest fishing industry, in tonnage of catch; and clear superiority in the supporting areas of ocean sciences and marine technology. But somehow the nation's leadership failed to husband and promote these advantages. Rather, it seemed to spend our principal, letting our maritime momentum run down until we have achieved second-rate (or worse) status in many areas. At the same time, several foreign sea powers rapidly built up their competitive capabilities. They quickly learned how to compete with U. S. flag interests on and in the oceans of the world. Why should the United States be concerned at all about being a sea power? More often than not, those of us who work in the ocean constituency take this as an immutable article of faith. However, this may not be a value shared by communities that are less directly involved.

Effective national security capability and a strong system for its worldwide maintenance can certainly be considered an "article of faith" by the citizens of our nation. Naval forces are a major component of this system. Since indications are that more and more of our strategic deterrence will be placed in submarines, this role will grow. As the most powerful nation in the west, the United States is obligated to support the principal share of the effort in maintaining the world balance of military power on land and in the sea and air.

About eighty-five percent of United States trade, by dollar value, is carried by merchant shipping of all flags, yet we only carry about five percent of our trade, by tonnage, in our own ships. Therefore, there is a strong argument for having more direct control over this trade system. By placing these lifelines of our economic well-being in the hands of foreign shipping companies, we run the grave risk of falling victim to those who might wish to reduce or modify our economic impact on world trade. Recent instances of the Soviet Union's rate cutting (operating at a loss to themselves) on certain international shipping routes indicate that they would be happy to put other shippers out of business to achieve monopoly on those routes. Is this a situation that U.S. shippers should be in? I think not. A more vigorous and workable U.S. maritime (i.e., marine transportation) policy is needed as one key element of our national sea power.

The United States competes all over the world in marketing of its products. We have the scientific base, high-technology industries, and capital to do this. The multibillion dollar ocean industry is no exception. Whether in the sale of offshore drilling equipment or in offshore gas and oil operations, our ocean industry should have greatly increased support from our government in order to be competitive with foreign enterprises.

Of course, there are many more supporting arguments for the development of a stronger national sea power stance. But maintenance of our national security and world trade integrity should suffice to illustrate the point.

The governance of national sea power comprises two inter-related areas: ocean policy and ocean management (or policy implementation). Policy establishes what is to be done, who will do it, the priorities, and the allocation of resources. Management is the process by which the directions of the policy-making mechanisms are carried out. Note the term "the policy-making mechanism." It is not the need to have a national ocean policy that is critical, rather it is the need for an effective policy-making mechanism.

While many contemporary studies stress ocean management, it is hard to consider the question completely without dealing first with the primary question of policy.

Ocean policy can be defined as that framework of decisions that plan or map out an integrated national program for uses of the sea. A coherent national ocean policy establishes directions for the government in how it views its sea power goals in the world vis a vis the other nations of the world. It assists in maximizing benefits while minimizing risks of conflict.

With respect to the oceans (which cover 71 percent of the surface of the earth) there is no such continuity of policy, no integrated United States national ocean policy. To be sure, there are many micro policies that deal with components of national uses of the sea, but the integrating whole is lacking. Without it, you only have muddling through and the potential for duplication and waste of both effort and public monies.

What are the components of national sea power? For the purposes of this discussion, we can divide them into a non-exclusive list of seven activity categories:

1. Defense
2. Living resources
3. Non-living resources
4. Energy
5. Marine transportation
6. Recreation
7. Marine research

Each of these areas has its own non-governmental constituencies, a complement of involved federal agencies, and congressional infrastructure. But in and among the areas there is really very little crossover, coordination, and cooperation. No one is in charge.

Thus, while each ocean activity is promoted by well-meaning, hard-working, and often zealous professionals, the net result, in the sense of the real national interest, often can be zero. Each constituency operates to maximize its situation even though this sometimes is at the expense of another equally deserving program. This uncoordinated approach to ocean management is analogous to building a house without a foundation. The walls are the various component uses of the sea, but the foundation must be a national ocean policy—a policy that integrates, coordinates, and provides priorities for national uses of the sea. Without this central direction, the present trend of losing our effectiveness as a major sea power will continue.

The Present State of United States Sea Power

How do we look as a sea power today? A few examples should be cited to illustrate the dimensions of the basic problem.

The present United States navy, with about 460 ships, is the smallest fleet in terms of size since 1939. More than ten years ago (1968) there were 1000 ships in the fleet. The Pentagon claims that the smaller, more modern fleet is composed of units that are much more effective than are the more numerous vessels of the past. They are; but consider, they can only be in one place at any point in time. One remembers what the great proponent of air power, Alexander de Seversky said, "Quantity is a quality in itself." We are also told that we should not compare the more numerous (2000 vessels) Soviet navy units with our navy, since the two navies have differing missions. The United States navy is designed to help protect U.S. interests at sea and overseas, while the USSR navy is scaled to defeat those interests and to protect "Mother Russia." But it would be a comfort to be able to insure that sufficient U.S. naval units could be available anywhere in the world where we might counter Soviet units. At present, the U.S. Navy just does not have the assets to do this. During the Carter administration, the Navy's requested shipbuilding program was not supported. Therefore, one cannot be optimistic that the United States can avoid having the second-best navy in the world. And being second best in this league is like being number two in any game with only two players—you lose!

The situation in our merchant marine is no more encouraging. In tonnage, the U.S. merchant fleet ranks ninth in the world after the United Kingdom, Norway, Japan, Liberia, Panama, Greece, the USSR, and France. Most of the major maritime states carry about 40 percent of their seaborne trade in their own flag vessels. Students of maritime affairs agree that this is a healthy ratio for maintenance of a strong national merchant marine. The United States flag fleet carries only four to six percent of its cargo tonnage. To be fair, if you separate out the tanker and bulk carrier percentages, then you find that our freighter fleet carries about 31 percent of U.S. trade in this class of cargo.

Despite the fact that the Merchant Marine Act of 1970 has done a great deal toward helping the U.S. merchant fleet to improve its assets, there are still real problems in the industry today. For example, the late 1970s saw the end of the U.S. flag passenger vessel operations and a reduction in the number of U.S. shipping companies. One major company, the Pacific Far East line, was dissolved through bankruptcy proceedings. Others are in precarious financial shape. Add to this the problem of rate-cutting by some heavily subsidized foreign shipping companies (notably those of the Russians) and you find that the U.S. merchant fleet faces even more problems in the future.

The notion that a free-enterprise, unsubsidized U.S. merchant fleet can compete in a substantially subsidized world maritime shipping marketplace is tenuous. Losing this trade for U.S. ships simply transfers the revenues for carriage of our goods to foreign governments via those nations' shipping companies. To be competitive, the subsidy structure for the U.S. merchant fleet must aim at keeping in the United States economy the monies now spent to support foreign merchant shipping companies.

In non-living resources, we have seen U.S. technological capability set world standards for the exploitation of marine minerals. But we have also seen the combined political process at all levels stymie the actual development of resources such as coastal gas and oil and the manganese nodules from the deep ocean floor. Lack of government leadership has prevented reasonable compromise between the interests of industry, our economy, and effective environmental management. Years and millions of dollars have been lost in fruitless dialogues while foreign competition has closed the gap on the lead we once enjoyed.

In fisheries tonnage, the United States ranks about sixth in the world, but in reality we take only about five percent of the world catch. To a large extent, the U.S. fisheries industry has been undercapitalized, unproductive, and obsolescent. It is estimated that fifteen percent of the world fish catch could come from waters within 200 miles of our coastlines. Thus, there is considerable potential for the United States fisherman to become a net exporter of fish if government were to develop some positive programs to help increase his productivity. In fact, we now *import* about $2.5 billion worth of fish a year! The U.S. Fisheries

Conservation and Management Act of 1976 appears to be having a positive effect in encouraging more fishing by U.S. interests. Perhaps this segment of our maritime industry is beginning to turn around.

Aquaculture, potentially an important supplementary source of fish production, suffers in the United States from lack of a relevant national development plan and the designation of a federal lead agency to guide it. Yet today, more than ten percent of the world's fish production comes from aquaculture. However, in the United States the figure is between one and two percent. The potential of aquaculture as a future food option for the United States is impressive. The lack of progress by the government in helping to promote a national effort in this area is therefore disappointing.

Energy from the oceans is another area where U.S. technology, managerial skills, and capital have led the way. In the mid-1960s most of the drill rigs in the world were owned by companies within 600 miles of Houston, Texas. Today, the U.S. offshore oil and gas industry meets stiff competition everywhere and in every fashion from several foreign nations. A visit to the foreign exhibits at the annual Offshore Technology Conference in Houston, Texas, will more than confirm this observation. And much of this foreign competition operates with the active support and encouragement of the respective governments. This is hardly the case in the United States. Alternative ocean energy schemes such as ocean thermal energy conversion (OTEC), tidal hydroelectric, wave forces, ocean currents, biomass conversion, and salinity pressure differential are possible future opportunities yet to be proved. They can offer promising energy development directions for U.S. technology if we can be assured of government support for the necessary basic research. There is a legitimate role for government in these areas, considering the uncertainty and enormous investment required for this research. The marine sciences is another area where there has been little real growth. To be sure, the figures used for "The Federal Ocean Program" reflect a funding growth from about $35 million in 1958 to well over $1 billion at present. It should be noted that the way the government adds up these figures has also changed. This is only a rough index of growth. But we seem to have added to the bureaucracy rather than to the programs that actually do the science. For many years, important basic and applied science programs such as those supported by the National Science Foundation, the Office of Naval Research, and Sea Grant have, on the average, grown at rates substantially less than the real inflation rate. At the same time, the nature of marine sciences has changed considerably. We recognize the importance of doing large-scale, long-time-period studies such as those represented by the International Decade of Ocean Exploration program of the 1970s. Exotic research platforms such as spacecraft, aircraft, and submersibles are being used more extensively in oceanography. These are expensive systems, but they now permit the conduct of research that could not be done just a few years ago. In many areas of marine research, costs have risen much more rapidly than has the national inflation rate. Ship operations is one of these areas. In addition, about one half of the present U.S. research fleet of 60 ships will require replacement within the next ten years. Despite this, through the several presidential administrations since the early 1970s, plans have been made for only one new research ship during the last decade—a 124-foot coastal research ship that is being funded by the National Science Foundation. It will join the university research fleet in 1981. I do not count the three vessels built in the mid-1970s for Oregon State University, the University of Rhode Island, and Woods Hole Oceanographic Institution, since these were directed by Congress rather than planned by presidential administration. After all, we are talking here about needs for ordering policy, planning, and management, not the exercise of power politics.

The lack of significant real growth in the marine science effort in the United States has had a double impact. First, it reduces the rate at which we increase our knowledge of the oceans and slows development of full potential economic value for the United States. Second, lower levels of effort constrain the educational system through which future scientific and technical manpower is developed.

Add to all of this the current "freedom of marine research" problems found in the ongoing United Nations Law of the Sea negotiations and you have a rather dim prospect for the immediate improvement of this area in the United States. It appears inevitable that we will have to live with a new order in the conduct of marine research within 200 miles of foreign coastal states. The final result will be an added cost burden for future marine scientists and built-in delay in the completion of their research tasks.

Marine recreation is a unique feature of a leisure-time, high-disposable-income society. Only a few of the coastal states in the world now have significant activities in this area. The United States clearly leads with an estimated $10 billion a year marine recreation industry. However, the growth of this industry has been accompanied by the very difficult problems of adjudicating complex, competing needs for the use of space along our crowded coastlines. Coastal zone legislation has been implemented at federal and state levels to deal with these problems. Useful solutions may be reached before the problems get out of hand.

There is, however, little appreciation of the real complexity of the process. It is exceedingly slow, and involves the active participation and cooperation of each of our coastal states in partnership with the federal government, citizens groups, industry, and local government.

What is the point of all this? Certainly nothing more can be done here than to give some brief examples, in a few areas, of how serious the total picture appears to be at this time. Essentially, the reason is that one must have an appreciation for the state of the integrated whole before considering individual remedies.

The Difficulties in Creating a National Ocean Policy

Why do we not have a national ocean policy mechanism and an effective management framework? There are several possible reasons:

1. The oceans do not and cannot offer a focused target for public and political attention. They are amorphous and diffuse environments. Their sheer scale and remoteness are not readily apparent to a man standing on the shore looking at only one edge of the sea. It is very unlikely that we will see either a "moon program" or a "Sputnik" event in the oceans. It would be hard now to imagine the kind of nice round crisis event that would trigger that sort of response for our national ocean program.

2. Ocean issues tend to be complex and interconnected. Few people have the training or experience to deal with a problem area where there is an interdisciplinary working environment of science, technology, economics, and politics.

3. Problems in the ocean environment tend to be insidious rather than sudden. Thus, the perception of real difficulties is usually achieved when it is too late to take any effective remedial action.

4. There is a real difference in "time constants" betweeen our operational government process and the time required to deal effectively with ocean problems. Ocean problems which require tens of years to solve, are being considered by people whose effective government service can be measured in terms of tens of months.

Expanding a bit on this last point can perhaps lead us to the heart of the problem. There is a sort of "governmental Gresham's Law" which says that short-term programs will displace long-term programs away from the attention of elected officials and their appointees. This is understandable. Presidential appointees have an average service life of about eighteen to twenty months. For example, even though President Carter asked his appointees to sign on for the full four years, his administration was not significantly different from others with respect to job tenure of its appointees.

But what can a man do in less than two years? He wants to have something to show for his short time in the government, so he selects short-term projects that will demonstrate that he is a real doer. Ocean projects that will not begin to show any results for five or ten years and which cost a lot, in his view, just will not be very attractive. Even the people who stay for the full four years really do not get that much time on the job. The presidential race begins early in the fourth year of office and many senior appointees are actively involved in the reelection campaign. These are the same people that we need to make the continuing basic policy decisions that are necessary to run the government.

Even if the same party gets reelected, there is a hiatus between the November elections and the installation of the new government the following January. At this point there is still a lot of job shuffling left to be done as old appointees are replaced by new. Thus, we probably can say that even a reelected administration is really effective for only about three years.

The situation is no better in the legislative branch, especially in the House of Representatives. Operating on a two-year cycle, a congressman, if he is junior or is not from a safe district, really gets about twelve or fourteen months of effective work on the job. Essentially, he must seriously think about reelection early in his second year. If he is successful, he still loses the time between November and January while he waits for the new Congress to convene. At that point there is added delay in getting on with the job as committee assignments are made and the business of Congress gets back up to speed. Senators, with their six-year terms, probably are in the best situation of any policy-level senior officials in the entire government.

This is not intended to be a short course in government, nor does it overlook the continuity provided by our Civil Service system. However, almost all of these people are below the real policy-making level.

The point here is that there is a serious mismatch between the half-life of the senior government *policy-makers* and the time needed to deal effectively with national ocean policy problems. We inherently lack the continuity necessary to solve those problems effectively. But there is no simple solution for this dilemma. The checks and balances of our governmental and political processes will not be altered on this account. Also, the long-term nature of ocean problems is not unique in our society. Similar difficulties could be found in many other long-term areas such as railroads, energy, and the environment, among others.

Centralized and Functional Views of Government Organization

Up to this point, we have dealt with the relationship between ocean policy and ocean management, the nature of national sea power and where some of the problems lie, and the fundamental conflicts between the operation of the government and the solution of ocean problems. Now, let us consider some proposed solutions for the establishment of a proper policy and management framework.

Ocean *management* within the federal government should be considered first. It permits derivation of the *policy* solution a little more easily.

The polar points of argument in the current ocean management dialogue appear to be either advocacy of a monolithic super ocean agency at the cabinet level, or nothing. In my view, neither of these approaches is acceptable; a reasonable solution lies somewhere in between. As the title of this chapter indicates, there is centralization and there is functionalization. Some might add "decentralization," but this term is really not applicable here.

The prevailing view in Washington, whether it is from the Commerce Department's Ocean Policy Study, the recent President's Reorganization Project, or the ongoing U.S. Senate National Ocean Policy Study is that we must create new centralized ocean organizations at some new order of scale. The scale ranges from creating cabinet-level agencies to consolidations within existing agencies. Parenthetically, I might add that most of these plans remind me a bit of an ice cream sundae; the obligatory whipped cream and cherry popped on top is usually the U.S. Coast Guard. They rarely get left out of any of the other agency plans! All this reflects an exercise in "bureaucracy in extremis." That is: if all else fails, reorganize.

There seems to be a general acceptance, *a priori*, that new organization is obvious and mandated. I disagree strongly, and propose that we consider instead rationalizing the ocean functions carried out by the various agencies within the government.

The existing basic design of the executive branch of the government is quite logical. There are departments for key functional areas such as food, natural resources, defense, money, health, education, energy, and so forth. The list below illustrates the point.

Executive Branch Functional Areas

Functional Areas	*Primary Federal Department/Agency*
Security	Department of Defense
International relations	Department of State
Food	Department of Agriculture
Money	Department of the Treasury
Transportation	Department of Transportation
Natural resources	Department of the Interior
Environment	Environmental Protection Agency
Commerce and trade	Department of Commerce
Health	Department of Health and Human Services
Education	Department of Education
Housing	Department of Housing and Urban Development
Labor	Department of Labor
Public Safety	Department of Justice
Compliance with the law	Department of Justice
Science and technology	National Science Foundation; National Aeronautics and Space Administration

Each of the major functional areas is under the direction of a principal (i.e., secretary or administrator) who reports directly and regularly to the President. If we begin to look at ocean activities in this context, that is, considering the basic goals of each activity, we find that we can sort most of them into the appropriate functional departments. For example, fisheries and aquaculture are primarily food-production activities. They should be in the "food department." Naval activities should be (and are) in the Department of Defense. Marine transportation should be in the Department of Transportation, and so on.

These are but a few examples of the fundamental point that just because something is wet does not mean that it should be put into a "department of wet." If a department's functional activities involve some marine aspects, then leave them in that department. The unnatural separation of wet and dry activities of government runs counter to the basic functional structure of the executive. Many studies and reports have expressed concern about the number of different government agencies and offices that are involved in ocean activities. I tend to agree, but for perhaps a different reason. Too many of these activities have been conducted or duplicated in the wrong agencies from a functional point of view. The departments have been permitted to poach on each other's territory and to develop competitive ocean efforts without any effective governance from higher authority. The result is a profusion of like-sounding activities in several agencies. There is a substantial difference between an agency competing for the lead in some new ocean efforts and its being invited to act as a supporting activity for the proper lead agency.

A Plan for Organization

To be more specific, let me give you my "perfect world" scenario— how I think the federal government should be organized, functionally, for ocean management:

1. Navy: the Department of Defense
2. Non-living natural resources and non-food living resources: the Department of the Interior
3. Living resources for food: the Department of Agriculture
4. Marine transportation: the Department of Transportation
5. Marine recreation: the Department of the Interior (National Park Service)
6. Coastal and offshore regulations, standards development, and enforcement: the Coast Guard
7. Ocean energy, with the exception of gas and oil natural resources: the Department of Energy
8. Environmental quality and protection: the Environmental Protection Agency

This is a bit of a coarse cut at the problem, but it supports my basic theme of simplification, reduction of overlap, and functionalization. Certainly there are problems where some areas do not overlap, but on balance this is the easiest direction to go. It involves a minimum amount of new statutory authority, if any, and thus it could be done largely within the organizational powers of the President. This is important, since any reorganization plan must be "do-able." A theoretically fine reorganization plan which does not take into account the political realities of dealing with the Congress and the affected constituencies is no plan at all.

Another thing to be considered in the superagency concept is whether or not it is effective to put virtually all of the federal ocean authority in the hands of one person. If the President makes a bad appointment then the whole ocean program could be adversely affected. On the other hand, it is unlikely that the President could make all bad appointments if ocean responsibilities were spread among several agencies on the basis that I have proposed here. Like the military principle of dispersion of forces in wartime, having several key presidential appointees instead of one "czar" would minimize "direct hits" in case of bad choices of key people.

The one major element that I cannot put neatly into my rational framework is the Army Corps of Engineers Civil Works activities that deal with navigable waterways and coastal areas. Separation of these functions from the Corps of Engineers would be difficult, and certainly there is no case for suggesting movement of the entire civil works function to another agency. Thus, I propose that they stay in the Department of Defense. But I am satisfied that my rather simple scheme, as outlined here, does give us the best arrangement.

Two new organizational elements would be created. The first would be a separate administration for the U.S. Coast Guard which would include the National Oceanic and Atmospheric Administration (NOAA) Commissioned Corps. In the case of the Coast Guard, I feel that its expanding regulatory and police functions call for its being an independent agency. Having the Coast Guard within a department of government which promotes the marine activities that the Coast Guard regulates creates an inherent conflict of interest situation. This new agency should be an administration with the administrator reporting to the Executive Office of the President, much as the administrator of the National Aeronautics and Space Administration (NASA) does.

The NOAA Commissioned Corps is added to this new administration because I feel that the coastal mapping, charting, and research tasks that they perform as a seagoing agency are closely related to the responsibilities of the Coast Guard. I would keep them a separate but coordinated part of the new United States Coast Guard Administration.

The second new organization would be a National Science and Technology Administration, which would combine the National Science Foundation, NOAA, and NASA. Looking only at the ocean functions in this new agency, it would embody both the basic and applied areas of marine science and technology as well as related scientific services.

Ideally, the resulting organization eventually would be elevated to a cabinet-level department of the federal government. It should be emphasized that a national science organization such as this must be designed with great care to insure that basic research is properly insulated from applied sciences and science services. By *insulated* I mean institutional mechanisms that protect basic research funding from being siphoned off into near-term problem areas. This is a problem that is common in situations where basic and applied sciences are mixed in the same agency. The problems of today tend to be financed by monies originally budgeted for future basic research. The future often gets mortgaged in favor of solving immediate problems. Optimum design of any science management organization must take this fact into account.

The foregoing still leaves us with the present system of *pares inter pares* (e.g., equal-rank cabinet officers) with respect to who will be in charge to insure that all of the ocean activities are conducted in a coordinated, effective way. This now leads us to consider the question of the ocean-policy-making mechanism.

To be effective, ocean policy must be conceived at the highest possible level of government. This should, in theory, be the President. Since the principal ocean managers are already at the cabinet level, under my scheme, the policy directorate necessarily must be above them to effect the proper supervision.

Ideally, the establishment of a specific Office of Management and Budget for Ocean Affairs is most attractive. The existing Office of Management and Budget (OMB) serves the administration by implementing presidential policies through assigning responsibilities to the various federal departments, insuring compliance, and allocating resources to carry out those responsibilities. As a practical matter, a "wet OMB" could not be created because the interface problems with the present OMB structure would be too complex to overcome.

Therefore, a two-element solution is proposed for the creation of an ocean policy-making mechanism. The first element would be the establishment of a National Ocean Policy Council within the White House. This unit would act as the principal advisory body to the President on ocean issues. It would also create and present ocean policy options for his consideration and approval.

The second element would be a well-defined ocean policy directorate within the OMB which would organize, implement, and follow up on the presidential decisions. The two elements would be required to work closely to insure a tight, vertical, policy-making structure.

It can be seen at this point that the National Ocean Policy Council would be somewhat analogous in its operations to the National Security Council. This council would facilitate intercoordination among federal department heads and would create national ocean policies that could be clearly assigned to the appropriate agency or agencies for execution. Outside advisory groups and panels would be created to help support the policy-making process.

The role of the existing presidentially appointed National Advisory Committee on Oceans and Atmosphere (NACOA) in such a scheme would become particularly important. It would be a primary source of outside advice to the President's ocean policy directorate. I would see NACOA expanding its activities while maintaining its independent organizational status in filling this role.

While I have dealt primarily with what needs to be done within the executive branch, since this is where the ocean leadership must originate, I should comment briefly on the relationship of Congress to my "perfect world" scheme. Essentially, I see few major problems with Congress in the executive branch realignment that I have suggested. Certainly there will be far fewer institutional/political/organizational problems generated by what is proposed here than by the various new agency proposals (e.g., a "wet NASA") that have been suggested by others. I do not see any need to suggest significant changes in the oversight, authorization, or appropriation committee structures. This proposal has by far the least impact on and requires the least change from Congress compared to the other centralized plans.

Summary

Ocean policy and management are nonseparable. One without the other will result in very little remedy for the critical problems that the United States faces as a sea power.

Centralization of ocean activities within a new or existing agency of the federal government will not work. In fact, it will only add to the problem because ocean issues must follow functional lines which are aligned with the basic organization of the executive branch.

Instead of attempting to create new agencies which further confuse the basic functional organization of the executive branch, a return to the fundamentals is needed. An audit of federal ocean activities should be conducted. Then the appropriate functions should be sorted into the appropriate departments and maintained there.

Policy formation and direction must come from a presidential-level council. When presidential decisions are made in this area, the authority to assign the tasks and resources to the appropriate department or departments should be exercised through an ocean policy directorate within the OMB.

Institutions rather than individuals, and continuity rather than convenience must be the basis for any reorganization in our national ocean programs.

If the United States does not soon take prompt action in dealing with the twin issues of ocean policy and management, then we can count on becoming a second-rate sea power within the next decade.

Some Final Thoughts

A final, urgent note is in order with respect to needed research in ocean policy. One of the widely understood and basic preliminary steps in developing any new programmatic area is the concurrent conduct of appropriate research. In the ocean policy and management areas, the current levels of research support reflect the degree of national attention that has been given to the whole question of sea power. They are very small. If you look at the nearly $1 billion "federal ocean program," an estimated less than one tenth of one percent is devoted to doing the necessary concurrent policy and program analysis. Yet it is here that we sort out the options, evaluate competitive positions, and facilitate operative information for the policy-makers. Such an analysis can function as a type of early warning system to avoid error and excessive costs. This is standard procedure in other places in our society, but in the ocean area we just do not do it.

As noted earlier, committees of the Senate, the House, the Library of Congress, the Office of Technology Appraisal, the Department of Commerce, the Department of Defense, and other institutions have been conducting ocean policy studies at various levels of complexity. But it is very difficult for an agency that has an interest in the outcome to do a balanced study in this area.

It is only through the well-understood system of contract research with universities, think tanks, and organizations such as the National Research Council that a reasonable approximation of impartial research findings can be achieved.

Institutional advocacy among federal agencies is healthy and necessary, but it should not be seen as the sole source of decision-making research and information.

While there is an understandable urgency to get immediate solutions, I would caution moderation. We have waited several decades for a proper sea power policy framework. We can afford to go slow while outside studies look at all the facets of the issues. This is not an area that has been studied to death—talked to death, perhaps, but this is not the same thing. Let us exercise due caution in choosing our options. We will have to live with the results for a long time.

A final note. Some casual readers will say that I have only suggested maintenance of the *status quo*. They should read this chapter again. This is, in fact, a comprehensive reorganizational outline. It is designed to consider the best and most economical (in a political sense) means for doing away with the *status quo* in a way that will permit the United States to rapidly realize its full potential as a sea power.

ANALYSIS

CHAPTER 4
ORGANIZING FOR MARINE POLICY:
SOME VIEWS FROM ORGANIZATION THEORY

STUART A. ROSS*

For at least two decades there has been a continuing controversy over how to organize the federal government's marine activities. There have been several changes, most notably the establishment of the National Oceanic and Atmospheric Administration, but there has been little sense of overall resolution. I do not propose yet another reorganization scheme; rather, this is an analysis of the problem of reorganization itself, drawing from a variety of materials in public administration and organization theory. How might reorganization problems be conceptualized? If this discussion helps those persons who are initiating proposals and those who are evaluating proposals, this chapter will have achieved its objective.

The discussion is limited primarily to ways of thinking about "packaging" a multitude of agencies and functions. Which functions should be attached to which agencies? Which agencies should be merged or split apart? Which agencies should be placed high or low in the hierarchy? "Packaging" is, of course, not the only important problem in marine policy, but it is a frequently discussed one.

There is no single theory or approach that allows satisfactory specification of the most appropriate way to reorganize government agencies. But neither is one reduced to blind experimentation. There is a range of useful, though partial, orientations. Each has its own considerable strengths and limitations; by reviewing them in comparison with one another it is possible to gain perspective on the ways in which we might think through marine reorganization problems.

The Efficiency Approach

The earliest views of administrative theory saw an organization as a tool for achieving rationality and efficiency. By merely rearranging the division of labor appropriately, the argument goes, duplicative efforts can be eliminated, communication channels can be simplified, and political interruptions can be obviated. In this view, "efficiency is thus axiom

* Stuart A. Ross is Assistant Director of the Sea Grant Institutional Program at the University of Southern California.

number one in the value scale of administration. This brings administration into apparent conflict with the value scale of politics. . . ."[1]

The assumptions of the approach are these: there are discoverable principles that govern the efficient organization of work units; correct organization of the work units is sufficient to ensure efficiency; individual behavior can be controlled by the job description and the economic reward; and for any organization there is a best formal division of the work.

This doctrine is, at least formally, the basis for all federal reorganizations in that it is specified in the Reorganization Act of 1949, which still governs such matters. The act instructs the President to:

> Examine and from time to time reexamine the organization of all agencies of the Government and . . . determine what changes therein are necessary to accomplish the following purposes;
>
> 1. To promote the better execution of the laws, the more effective management of the executive branch of the government and of its agencies and functions, and the expeditious administration of the public business;
>
> 2. To reduce expenditures and promote economy, to the fullest extent consistent with the efficient operation of the Government;
>
> 3. To increase the efficiency of the operations of the Government to the fullest extent practicable;
>
> 4. To group, coordinate, and consolidate agencies and functions of the Government, as nearly as may be, according to major purposes;
>
> 5. To reduce the number of agencies by consolidating those having similar functions under a single head, and to abolish such agencies or functions thereof as may not be necessary for the efficient conduct of the Government; and
>
> 6. To eliminate overlapping and duplication of effort.[2]

The debates over organization of marine affairs that emerged in the 1950s and 1960s have included several versions of this doctrine of efficiency. One emphasis has been on agency cooperation to ensure the

---

[1] Luther Gulick, quoted in Harold Seidman, *Politics, Position, and Power: The Dynamics of Federal Organization* (New York: Oxford University Press, 1975), p. 6. See also the following works: Frederick Winslow Taylor, *The Principles of Scientific Management* (1911. Reprint, New York: W. W. Norton, 1967); Luther Gulick and L. Urwick, eds., *Papers on the Science of Administration* (New York: Institute of Public Administration, 1937); and Henri Fayol, *General and Industrial Management*, trans. Constance Storrs (Pitman, 1949).

[2] Quoted in Seidman, *Politics, Position, and Power*, pp. 9-10.

efficient use of marine facilities, manpower, and data. The Interagency Committee on Oceanography, in operation from 1960 to 1966, was successful in several separate efforts to this end. Another emphasis has been the attempt to group not just operational efforts by agencies but agencies themselves so that greater efficiency can be obtained from putting similar operations together on a continuing basis. This effort continues to the present, down to the debates between supporters of President Carter's proposal for a Department of Natural Resources and supporters of NACOA's proposal for an independent agency.

The efficient management approach has been heavily criticized by those who have studied organizations as being inapplicable to all but the most narrowly defined problems.

First, many authors have objected to the approach on the grounds that human beings who work in organizations are not guided by the simplistic rationality presumed of them by management designers of this philosophy. Workers do not always respond positively to monetary or other incentives that seem (to outsiders) to be in the workers' best interests; they have emotions, psychological needs, and group attachments as well.

Second, it needs to be remembered that the organization is not an isolated mechanism that is subject only to the actions of the designer or manager. It is subject to many forces from its social environment—forces such as social traditions, price changes, technological advances, and so on. The experience of the Bureau of Land Management, which had major marine programs given to it and then taken away from it in the 1970s, suggests that external forces can be of considerable importance.

Third, the efficiency doctrine often mistakenly assumes a single purpose for the organization, a single scale against which accomplishment can be measured. In most organizations there is, instead, a diversity of goals and of perceptions about the world. If one goal is stated, it is often broad and ambiguous, permitting several different interpretations in practice, and there are usually several managers helping to implement the goal, each with a different personal interpretation in mind. Try, for example, to state a single precise purpose for the Coast Guard, which intercepts smugglers, enforces fishing laws, and blocks oil spills (and more).

This diversity means that assertions about efficiency of performance may be open to wide dispute. If we are in disagreement about whether an agency's role is to encourage conservation or to encourage development, to encourage technical accuracy or to encourage public participation, or to do all of them, then how will we know which structural alternatives would be most efficient? If we are at odds over how to evaluate good internal communications, then how will we arrange the communications pathways?[3]

Further, this diversity of goals and perceptions means that attempts to link similar or interdependent activities together will also be subject to wide disagreement. We are accustomed by now to finding that persons with strong ties to land-based activities see marine activities as subsets of their own work, whereas those in marine activities often proclaim the interdependence of all things marine. The perception of interdependence also varies with resources: he who has enough resources to pay off or ignore affected parties sees or feels little interdependence between his actions and those of other parties. Finally, the perception of what things belong together is a product of social conventions: energy politics and marine politics, both unsettled now, may in a few decades appear rather conventionalized and confined, as labor and agriculture do to most of us now.[4]

As Ernst Haas says, there "is no clear-cut logic which links the characteristics of problems with appropriate and known organizations."[5] The appropriate organization pattern depends also on the purposes of the organizer, i.e., what further interdependencies he wishes to create. Haas argues that the appropriate international organizational pattern in response to the perceived interdependencies of ocean activities depends on whether one intends technical improvements, broad economic development, or world government.

For all these reasons, many scholars and managers have moved on to other approaches.

The Political Approach

Another familiar view of bureaucracies is that they are basically political arenas in which causes are won or lost and careers are advanced

---

[3] For good critiques of the "proverbs of administration," see Herbert A. Simon, *Administrative Behavior*, 2nd ed. (1957. Reprint New York: The Free Press, 1965) and Herbert A. Kaufman, "Reflections on Administrative Reorganization," in *Current Issues in Public Administration*, ed. Frederick S. Lane (New York: St. Martin's Press, 1978), pp. 214-233.

[4] Ernst B. Haas, "Is there a Hole in the Whole: Knowledge, Technology, Interdependence, and the Construction of International Regimes," *International Organization* 29 (Summer 1975), pp. 827-876. See also the analysis by Elliot Richardson, in testimony before Congress, cited in U.S. Department of Commerce, *U. S. Ocean Policy in the 1970s: Status and Issues*, p. IX-27.

[5] Haas, "Is There a Hole in the Whole," p. 848.

or stalled in incessant rounds of bargaining and scheming. In particular, organizational structures are seen as the result of bargaining and compromise among persons or groups that have different motives and different amounts of power. The establishment of a new department, the formation of a coalition, or any other structural development is to be explained by reference to the factional struggles that preceded it or to the intentions of those who put it together. Thus we hear it argued that NOAA went to the Department of Commerce because President Nixon disliked the Secretary of the Interior, Walter Hickel,[6] or that President Carter dropped his plans for a Department of Natural Resources because he could not afford the predictable battles in congress.

These theories assume that people are driven by either personal advancement or attachment to some particular policy, rather than by the abstract idea of efficiency. It also assumes that the individual actors possess various kinds of resources such as money, information, personal contacts, or official positions, that can be applied to the pursuit of their goals. Finally, it is assumed that individual actors are more or less rational in their pursuit of their objectives: they know what they want and act accordingly when confronted with particular rewards or opportunities. Changes in the bureaucracy are presumed to be in somebody's self interest, and they occur because that somebody had better calculations or more resources, or both.

A very popular statement of the political perspective is Graham Allison's "Model III," one of three perspectives which he reviews in showing the many faces of a complex policy situation—the Cuban missile crisis.[7] In Allison's model, the political players—persons and groups of persons—have different amounts of power and sources of power. The players tend to have parochial viewpoints; for any given issue, different players will see different aspects of it. The bargaining usually takes place along recognized "action channels" such as the budgeting process, implementation of a command down the hierarchy, regular consultation and clearance circuits, and the like. Furthermore, the political player is under pressure; he is always facing deadlines, and he always has competitors waiting for him to slip. The formally designated leaders do not automatically play dominant roles; their formal positions give them special resources, but otherwise they are players like everyone else. Outcomes, including structures, are to be understood not as the solutions to problems, but as political resultants:

---

[6] *Ocean Science News*, July 30, 1979, p. 6.

[7] Graham T. Allison, *Essence of Decision: Explaining the Cuban Missile Crisis* (Boston: Little Brown, 1971).

> . . .*resultants* in the sense that what happens results from compromise, conflict and confusion of officials with diverse interests and unequal influence; *political* in the sense that the activity from which decisions and actions emerge is best characterized as bargaining along regularized channels among individual members of the government.[8]

Allison's work captures the spirit of the political view quite well, but it is not very precise as a theoretical formulation, being somewhat ambiguous on such matters as the degree of rationality one should attribute to the actors and whether individuals or groups are the relevant units of analysis.

Gordon Tullock gives special emphasis to rational behavior by individuals; his subject is the "intelligent, ambitious, and somewhat unscrupulous man in an organization hierarchy."[9] Tullock makes many of the economist's assumptions about man (rationality and utility maximization, for example), but he assumes an environment very unlike a market, namely a hierarchy in which subordinates are such largely because they have few meaningful alternatives to their present positions. (Middle-level executives, Tullock argues, often exhibit more servility to their superiors than a common laborer ever would.) The organizational politician lives in a world that is divided simply into sovereigns, subordinates, and equals. The organizational politician is influenced by more general forces as well—unconscious cultural influences, ethical rules, organizational patriotism, pressures toward proper social behavior, and promotion criteria. Many familiar structural matters, such as monitoring procedures, communications networks, career paths, coordination mechanisms, empire building, and so forth, can then be seen in terms of the attempts by subordinates to rise and by sovereigns to control. Reforms can be (and are) put forth that seek to tailor organization structures to the facts of human behavior as Tullock sees them. A less self-consciously theoretical work that also emphasizes individual politics is Dalton's detailed study of the informal power plays of the executives in a factory.[10]

Another popular variant of the political view focuses on the agency rather than on the individual as the relevant actor. Francis Rourke, for example, asserts that bureaucracies are necessarily political, and that power differences between agencies stem from one or more of the following characteristics: the agency's constituency, its type and amount

---

[8] Allison, *Essence of Decision*, p. 162.

[9] Gordon Tullock, *The Politics of Bureaucracy* (Washington, D.C.: Public Affairs Press, 1965), p. 26.

[10] Melville Dalton, *Men Who Manage* (New York: John Wiley and Sons, 1959).

of expertise, its esprit, and its leadership.[11] Unlike Tullock, Rourke is not self-consciously theoretical; unlike Allison, he does not give much consideration to power plays between individuals. He treats the agency almost as if it had a single mind of its own.[12] Rourke leaves moot the questions of whether the agency seeks power for self-aggrandizement or for substantive accomplishment.

In the organizational politics perspective, structures are resultants; but they are also the givens with which the next analysis begins. Structures do emerge as the state of affairs after the battle has been fought: the list of units assigned to NOAA and the jurisdictions of Congressional committees, for example, may both be seen in this light. But no political battle can be described without specifying the structures within which it takes place. To understand the battle, one has to know at the outset that there were, say, three points of view, and two departments, and so much money on each side. Each of the examples just given has served or will serve this function as one of the parameters in later political battles.

The political view is obviously employed quite frequently in analyses of the marine reorganization problem. For example, consider this analysis:

> The debate on the form that a new ocean organization should take often turns on how much influence or "clout" an independent agency could muster as compared to a Cabinet-level department. Often, the influence depends less on the rank of the administrator than on his personal influence in the Administration. However, it is generally believed that cabinet level officers have greater access to and are more influential with the President and other Cabinet officers than are administrators of independent agencies or sub-Cabinet-level administrators.[13]

The marine newsletters and magazines, such as *Ocean Science News* and *Coastal Zone Management*, rely heavily on the political view; so do conversations at marine-related conventions; and so do many historical works such as Edward Wenk's *The Politics of the Oceans*.[14]

This perspective incorporates a sense of time and history much more directly and with less self-conscious effort than do most of the other approaches. The political view presents a moving record of which participants won which battles and attained which heights. Each result is

---

[11] Francis E. Rourke, *Bureaucracy, Politics, and Public Policy* (Boston: Little, Brown, 1969).

[12] For another treatment of this sort, see Matthew Holden, "'Imperialism' in Bureaucracy," *American Political Science Review*, 60 (1966): 943-951.

[13] United States Department of Commerce, *U. S. Ocean Policy in the 1970s: Status and Issues*, IX-24.

[14] Edward Wenk, *The Politics of the Oceans* (Seattle: University of Washington Press, 1972).

the given for the next stage. Individual careers can be followed, and so can the rise and fall of various political groupings. An organizational history written in this view is fed by two streams which are not generally explored further: the appearance of new participants, events, and other phenomena; and the drives or motives of the various participants.[15]

One major and well-known problem with this approach is the difficulty of defining the concept of power. Is power the control over the uncertainty experienced by others, the control over particular resources, the ability to influence someone else's behavior, or some other aspect of the relationship between two persons?[16] Scholars have been unable to agree on a definition that is both theoretically and empirically useful.

Furthermore, hard evidence about personal power relations is almost always difficult to gather, for these matters are not often discussed in official documents, are not readily captured by numerical data, and are subject to social and legal strictures on discussion. There is often an air of gossip and illegitimacy about such information.[17]

Thus arises the most serious problem with the political view of organizations. Unless the motives of the participants and the definition of power can be specified in advance, and in the absence of detailed evidence, the political view simply becomes non-falsifiable. Motives and sources of power are merely inferred after the fact to explain whatever situations came up. If a person gets what he wanted, he must have wanted to, and he must have had more power than the others.

The Humanistic Approach

One reaction to the rational-efficient view of organizations has been that it ignores or distorts the psychological and social strengths and limitations of the human beings who make up an organization. Even the political view, although it emphasizes individual differences, ignores many

[15] For more on this point see Andrew W. Pettigrew, *The Politics of Organizational Decision-Making* (London: Tavistock Publications, 1973), especially chapter 1.

[16] For these different views of power, see, respectively, Michel Crozier, *The Bureaucratic Phenomenon* (Chicago: University of Chicago Press, 1964); Francis E. Rourke, *Bureaucracy, Politics, and Public Policy*; and James G. March, "An Introduction to the Theory and Measurement of Influence," *American Political Science Review* 49 (1955):432-451.

[17] Gerald R. Salancik and Jeffrey Pfeffer quoted Warren Bennis as calling politics the organization's "last dirty secret," in their "Who Gets Power—and How They Hold on to It: A Strategic Contingency Model of Power," *Organization Dynamics* 5 (Winter 1977), pp. 3-21. See also Bronston T. Mayes and Robert W. Allen, "Toward a Definition of Organizational Politics," *Academy of Management Review* 2 (October 1977): 672-677, which defines organizational politics as the use of illegitimate means for legitimate ends or vice versa.

emotional and social factors. Attention must also be given to social groups, informal work norms, personality needs, and styles of communication. Organization structures put together without attention to such factors may prove ineffective, whatever their other merits.

Although it seems to be simply common sense now to say that personal and group relations exist on the job in patterns that are different from the formally designated patterns, the lesson did not come easily to early administrative theorists. The lengthy Hawthorne experiments in the 1930s jarred scholars into an appreciation of psychological and social factors, and case studies by Blau and others added to the evidence of the importance of such factors.[18]

Developing from these studies has been a line of thought that management styles and structures should relate to such "extra-rational" needs of persons and groups—on the grounds that so doing will improve organizational performance.

In the works of Chris Argyris, for example, it is theorized that if the organization is such that it supports and develops the personality, psychological energy will be put into organizational performance.[19] If the organization frustrates the personality, however, psychological energies either will not be forthcoming or will be directed into channels that are dysfunctional for the organization: frustration, conflict, cheating, and so on.

Another author, Rensis Likert, reported research indicating that greater productivity is attained if management emphasizes supportive personal relations, group-based decision-making, and high performance goals.[20] Likert posited a spectrum of management styles, from "exploitative-authoritative" through to "participative group" (later changed to less evaluative names, from "System 1" to "System 4"). His survey data indicate that organizations that emphasize the three concepts of supportive relations, group-based decisions, and high performance goals are more productive than organizations that do not. Although Likert relies more on quantitative results and less on elaborations of personality theory than Argyris does, he obviously shares the general normative thrust of Argyris.

[18] The Hawthorne experiments are reported in: F. L. Roethlisberger and W. J. Dixon, *Management and the Worker* (Cambridge, Mass.: Harvard University Press, 1939); George C. Homans, *The Human Group* (New York: Harcourt Brace, 1950); and Elton Mayo, *Human Problems of Industrial Civilization* (Boston: Graduate School of Public Administration, Harvard University, 1946). See also Peter M. Blau, *Dynamics of Bureaucracy*, rev. ed. (Chicago and London: University of Chicago Press, 1963).

[19] Chris Argyris, *Personality and Organization* (New York: Harper and Row, 1957); and *Integrating the Individual and the Organization* (New York: Wiley, 1964).

[20] Rensis Likert, *New Patterns of Management* (New York: McGraw-Hill, 1961); and *The Human Organization: Its Management and Value* (New York: McGraw-Hill, 1967).

Another study portrays the social structure of the organization as having these aspects: assigned roles, internal interest groups, social stratification, belief systems, patterns of commitment and participation, and patterns of dependency. In *Leadership and Administration*, Philip Selznick argues that organization structures are based in persons and their relations to one another; he identifies *leadership* as the most important independent variable affecting the organization.[21]

The key tasks of leadership include "the definition of mission and role," "the institutional embodiment of purpose," "the defense of institutional integrity," and "the ordering of internal conflict." Effective leadership transforms a mere organization, "a rational instrument engineered to do a job," into an institution, "a natural product of social needs and pressures," "infused with value." [22]

Two other related works may be mentioned briefly. One such study distinguishes between "Theory X" management and "Theory Y" management. Theory X assumes that people want to avoid work, that they require close supervision, and that they do not seek responsibility; Theory Y assumes that people will seek work and responsibility and that they are not necessarily in conflict with the organization. [23] It turns out, in this study, that Theory Y works better. Other authors propose a "grid" for describing managers in terms of two characteristics: their emphasis on production and their emphasis on people. The best managers score high in both. [24]

These works do not pretend to offer any solutions about where to place the formal boundaries between work units, but they do argue implicitly or explicitly that no formal structure can be satisfactory unless an appropriate management style is used.

Another group of studies on the humanistic tradition does argue the actual division of labor, saying that the particular jobs to be done in and by the organization should be designed in accordance with the needs and potentials of the human personality.[25] Although the research designs of

[21] Philip Selznick, *Leadership in Administration* (Evanston, Ill.: Row, Peterson, and Co., 1957).

[22] Selznick, *Leadership in Administration*, pp. 5, 40.

[23] Douglas McGregor, *The Human Side of Enterprise* (New York: McGraw-Hill, 1960). See also Arthur Kuriloff, "An Experiment in Management: Putting Theory Y to the Test," *Personnel* 40 (November-December, 1963): 8-17.

[24] Robert Blake and Jane Mouton, *The Managerial Grid* (Houston: Gulf Publishing Co., 1964).

[25] See Lisl Klein, *New Forms of Work Organization* (Cambridge: Cambridge University Press, 1976); Ramon J. Aldag and Arthur P. Brief, *Task Design and Employee Motivation* (Glenview, Ill.: Scott, Foresman, & Co., 1979); and George Strauss, "Job Satisfaction, Motivation, and Job Redesign," in *Organizational Behavior: Research and Issues*, ed. George

these studies have sometimes been flawed and the data are impressionistic, the consensus has been that new ways of dividing up the work can significantly improve morale and productivity.[26] The general line of argument, which is now conventional wisdom with the educated public, is that repetitive and narrowly constructed jobs, such as those on a stereotypical assembly line, are inherently dehumanizing and are therefore ultimately detrimental to productivity. Larger and more complex roles for individual workers are usually recommended. For example, an automobile factory might decide to replace an assembly line with a process in which each car is assembled from start to finish by one group of workers.

Many newer studies have been put forth under the rubric of "organizational development"; "OD" is currently popular in management consultation efforts.[27] Although most of these more recent works have broader concerns than, say, Argyris or McGregor, the emphasis on individual and group behavior remains a major identifiable theme. OD also emphasizes action within the organization along with research.

Whatever their differences, the humanist studies share three important characteristics. First, they favor open, supportive, affective relations with and among workers. Second, they nonetheless retain the manager's emphasis on efficiency and performance—they differ with the earliest administrative theorists on what human nature is but not often on the appropriateness of guiding it toward one's own ends. Third, these studies (like the administrative theorists and unlike the political theorists) are usually ready to prescribe solutions for managerial problems.

Whether the formal and informal boundaries within an organization do or should coincide is an intriguing and still unresolved issue in these studies. Some writers assume that the formally assigned work group is and should be the focus for social relationships. One textbook on public administration, for example, concludes its discussions of the formal vs. informal problem with the observation that "the ultimate aim is to make the formal and the informal organization converge."[28] Writers such as

Strauss, Raymond E. Miles, Charles C. Snow, and Arnold S. Tannenbaum (Belmont, Calif.: Wadsworth, 1976), pp. 19-49.

[26] This point is made by Lisl Klein, in *New Forms of Work Organization*, chapters 4 and 5.

[27] For good treatments of organization development, see Robert T. Golembiewski, "Organization Development in Public Agencies: Perspectives on Theory and Practice," *Public Administration Review* 29 (July-August 1969): 367-377; Frank Friedlander and L. Dave Brown, "Organization Development," in *Annual Review of Psychology*, ed. Mark R. Rosenzweig and Lyman W. Porter (Palo Alto: Annual Reviews, Inc., 1974); and Wendell French, "Organization Development: Objectives, Assumptions, and Strategies," *California Management Review* 12(1969):23-34.

Likert and Homans take a similar view. They see an organization as made up of formally assigned work groups that are also socially cohesive internally. Other theorists, including those associated with the political view of organizations, assume that there are significant coalitions and connections across formal boundary lines, and that these may present real dilemmas for the organizational participants.

The humanist view has played a rather small part in the debates over marine reorganization. Presumably, individual marine agencies have had problems with management style and with personality needs, as do all other organizations; and presumably some of the agencies have been worked over by OD teams or studied by students of psychology or sociology. Moreover, the cultivation of cooperative and supportive relations between agencies would considerably mitigate the kind of jurisdictional conflicts that plague marine policy, as, for example, between marine and freshwater fisheries problems. Such suggestions for marine policy do not usually go beyond a few exhortatory paragraphs, however, perhaps because the case studies of the humanist scholars rarely take in so great a scope as an entire policy field, or perhaps because the impossibility of much success in the federal environment is a foregone conclusion.[29]

The Technology-Environment Approach

The three approaches just described usually treat only the organization itself, without much regard for the external forces, economic and social, that may be affecting it. Studies in the previous modes also tend to ignore any constraints imposed by the kind of work that the organization is trying to do. They often imply that a large government agency, a department store, and a political campaign organization may be viewed as equivalent structural problems. A newer body of work builds instead on the premise that organizations so differently situated may well have quite different structural requirements.

This fourth view of organization structure emphasizes therefore the importance of an organization's "environment" (social and economic settings) and its "technology" (types of work, work routines, and materials) in determining the organization's structure. An organization's environment and technology are thought to be important because they

---

[28] George E. Berkley, *The Craft of Public Administration*, 2nd ed. (Boston: Allyn and Bacon, 1978), p. 100.

[29] See, for example, U.S. Department of Commerce, *U.S. Ocean Policy in the 1970s: Status and Issues*, p. IX-23.

present informational or physical constraints or opportunities with which the organization seeks to cope by structuring itself accordingly. "The main burden of the . . . approach is to identify 'outside' or extra-administrative elements which limit the choice of effective control structures."[30] Useful summaries of these studies are those by Dessler and by Kast and Rosenzweig.[31]

For example, a widely repeated finding has been that organizations experiencing considerable complexity or change do or should have more flexible internal structures. [32] Another is the notion that organizations segment themselves to match the segmentation of their environment.[33] Whatever the particular findings, there is agreement that technological and environmental characteristics are useful independent variables for comparative analysis of structures among many organizations.

The technology-environment theorists often put forward prescriptions for organization design in conditional form: if the organization has Technology X and Environment Y, it should have Structure Z. The prescriptions are derived from findings that organizations with structures that match their situations are more successful.[34]

These theories usually conceive of the organization as a system, implying or stating that they see patterned dynamic relationships among the elements, exchanges with the environment, goals to be achieved, a hierarchy of internal subsystems, and transformations applied to inputs from the environment. The transformations or internal processes, which may also be understood as the means toward the system's ends, are the organization's technology.

---

[30] Christopher C. Hood, *The Limits of Administration* (London: John Wiley and Sons, 1976), p. 144.

[31] Gary Dessler, *Organization and Management: A Contingency Approach* (Englewood Cliffs, N.J.: Prentice-Hall, 1976); and Fremont E. Kast and James E. Rosenzweig, *Contingency Views of Organization and Management* (Chicago: Science Research Associates, 1973). These studies are sometimes collectively referred to as contingency theory, on the argument that the technology and the environment present contingencies with which the organization must cope. It should be noted, however, that use of the "contingency" idea has also been made by several authors who view organization structures largely in terms of persons and groups— an overlap between the third and fourth approaches arbitrarily delineated here. See Don Hellriegel and John W. Slocum, Jr., *Organizational Behavior: Contingency Views* (St. Paul, Minn.: West Publishing Co., 1976).

[32] Kast and Rosenzweig, *Contingency Views of Organization and Management*, part 3.

[33] See, for example, James D. Thompson, *Organizations in Action* (New York: McGraw-Hill, 1967), p. 67. These arguments are sometimes traced to W. Ross Ashby's "Law of Requisite Variety." See his *Introduction to Cybernetics* (New York: John Wiley and Sons, Inc., 1956), chapter 11.

[34] See, for example, Paul R. Lawrence and Jay W. Lorsch, *Organization and Environment* (Homewood, Ill.: Richard D. Irwin, 1969); and Joan Woodward, *Industrial Organization: Theory and Practice* (London: Oxford University Press, 1965).

The technology-environment management perspective is highly applicable to marine reorganization problems. It advises us that the rapid scientific and social changes expected in the decades ahead should be met with flexible and informal structures rather than with rigid machine-like structures. (Perhaps, therefore, we should not feel uncomfortable with the unsettled character of our marine programs.) It cautions us that the structures suitable for running a fleet of research ships may not be suitable for planning uses of the coastal zone. It advises us that the flexibility and variations we want for marine problems may be quite difficult to reconcile with the largely predetermined federal environment, complete with civil service rules and elaborate budgetary reviews constraining every agency. It justifies theoretically our practical experience that the organization of agencies depends on the organization of congressional committees, since the agencies are so strongly affected in practice by their legislative and appropriations committees. Similarly, it makes organizing around constituency groups seem an appropriate adaptation rather than a cop-out.

Several limitations of the contingency theories have become evident, however, including at least the following three (overlapping) points.

First, organization members and organizations react to the technology and the environment they perceive rather than to what the observer or researcher perceives.[35] Therefore, attempts to build an organization's structure according to "the" technology and environment may founder on personal and political disagreements about the very nature of the organization's work and of the world in which it operates. The efficiency approach encountered similar difficulties.

The concept of technology turns out to be elusive conceptually and operationally for other reasons as well. It can be confused with structure itself; for example, is "specialization" a characteristic of the work or of the way the work has been structured?[36] Further, talk about technology as the means toward the organization's goals presumes agreement on the goals, which in policy areas like marine policy may simply not exist. In one case, later studies found that the structural effects of technology in a large organization could not be detected beyond the workflow units

---

[35] See Charles Perrow, *Complex Organizations: A Critical Essay* (Glenview, Ill.: Scott, Foresman, 1972); Karl E. Weick, *The Social Psychology of Organizing* (Reading, Mass.: Addison-Wesley, 1969); and John Child, "Organizational Structure, Environment, and Performance: The Role of Strategic Choice," *Sociology* 6 (January 1972): 1-22.

[36] See Perrow, *Complex Organizations: A Critical Essay*; Gary Stanfield, "Technology and Organization Structure as Theoretical Categories," *Administrative Science Quarterly* 21 (1976): 489-493; and Charles A. Glisson, "Dependence of Technological Routinization on Structural Variables in Human Service Organizations," *Administrative Science Quarterly* 23 (September 1978): 383-395.

themselves, suggesting that the early findings of the importance of technology occurred only as a result of having studied smaller firms.[37]

The concept of environment is equally difficult for complex organizations, where interactions with the environment are varied and in some places stronger than many connections to the inside. In administrative or regulatory organizations, unlike factories, virtually every member is both a worker and a boundary-spanner to the environment. The environment-organization distinction is most clear at a great distance or for very formal purposes; yet most of the work done by, for, and to such agencies is done by people and organizations who are enmeshed closely and often informally.

As a result, the cumulation of studies has not yet gone much beyond a general agreement that uncertainty and change require flexible structures and a few notable attempts to specify internal substructures for particular contingencies.[38]

In short, it appears that the constraints imposed by technology and the environment can be and often are either overcome by other factors or obviated by other interpretations: the constraints of technology and environment turn out to be "plastic" rather than "iron," i.e., surmountable at a cost. [39]

The Social Construction Approach

In ᐟ this final perspective, emphasis is given to the view that organization structures arise not so much by implementation of a single conscious plan as by the cumulation of various social processes in which members and observers collectively negotiate and construct their own particular organizational worlds.

> Organizational realities are not external to human consciousness, out there waiting to be recorded. Instead, the world as humans know it is

[37] See Joan Woodward, *Industrial Organization: Theory and Practice*, and D. J. Hickson, D. S. Pugh, and Diana C. Pheysey, "Operations Technology and Organization Structure: An Empirical Reappraisal," *Administrative Science Quarterly* 14 (September 1969): 378-397.

[38] See James D. Thompson, *Organizations in Action*, and Jay Galbraith, *Designing Complex Organizations* (Reading, Mass.: Addison-Wesley, 1973) and Jay Galbraith, *Organization Design* (Reading, Mass.: Addison-Wesley, 1977). Both authors carefully delineate sets of rules or steps for building organizations in the face of contingencies. Thompson, for example, sees the containment of interdependencies and the coping with uncertainties as the driving force for structuring; he posits particular structural responses that are likely and rational in response to different kinds or levels of interdependence and uncertainty.

[39] The terminology is attributed to Karl Popper in Gabriel A. Almond and Stephen J. Genco, "Clouds, Clocks, and the Study of Politics," *World Politics* 29 (July 1977): 489-522.

constituted intersubjectively. The facts of this world . . . are neither subjective nor objective in the usual sense. Instead, they are construed through a process of symbolic interaction.[40]

Writers such as Weick, Ranson, Brown, and Benson typify this tradition.[41] Although these ideas are relatively new in the literature on organizational structures, they draw on substantial intellectual antecedents in phenomenology and social psychology.[42]

In this view, the reality of the organization's parts and relations is established by processes of individual perception and also by processes of social interaction—in current jargon, organizational situations are "enacted" in incessant social negotiations. It is only through such processes that groups of people are at all able to establish a shared sense of what is real and what is not. But the structures people see and to which they think they belong act, in turn, upon the people, influencing their behavior and their perceptions. Ranson argues that we must move to:

> . . .a way of seeing structures as a vehicle constructed to reflect and facilitate meanings. Structural frameworks systematically embody normative expectations and prescriptions for competent and satisfactory performance.[43]

In more conventional sociological terms, sets and boundaries in the social world can and do act as reference groups, supplying the individual with norms, expectations, rewards, and a sense of identity.[44] Of course,

[40] Richard Harvey Brown, "Bureaucracy as Praxis: Toward a Political Phenomenology of Formal Organizations," *Administrative Science Quarterly* 23 (September 1978): 365-382.

[41] Karl E. Weick, *The Social Psychology of Organizing*; Karl E. Weick, "Educational Organizations as Loosely Coupled Systems," *Administrative Science Quarterly* 21 (March 1976): 1-19; Karl E. Weick, "Organization Design: Organizations as Self-Designing Systems," *Organizational Dynamics*, Autumn 1977, pp. 31-46; Richard Harvey Brown, "Bureaucracy as Praxis," pp. 365-382; J. Kenneth Benson, "Organizations: A Dialectical View," *Administrative Science Quarterly* 22 (March 1977): 1-21; Stewart Ranson, Bob Hinings, and Royston Greenwood, "The Structuring of Organization Structures," *Administrative Science Quarterly* 25 (1980): 1-17.

[42] For discussion and reference see Bernard N. Meltzner, John W. Petras, and Larry T. Reynolds, *Symbolic Interactionism: Genesis, Varieties, and Criticism* (London: Routledge and Kegan Paul, 1975); Karl E. Weick, *The Social Psychology of Organizing*; Peter L. Berger and Thomas Luckmann, *The Social Construction of Reality* (Garden City, N.Y.: Doubleday and Co., 1967); Burkart Holzner, *Reality Construction in Society* (Cambridge, Mass.: Schenkman Publishing Co., 1972).

[43] Ranson, Hinings, and Greenwood, "The Structuring of Organizational Structures," p. 3.

[44] See, for example, Robert K. Merton, *Social Theory and Social Structure* (Glencoe: The Free Press, 1957), and H. H. Kelley, "Two Functions of Reference Groups," in *Readings in Social Psychology*, ed. G. E. Swanson, T. M. Newcomb, and E. L. Hartley, rev. ed. (New York: Holt, 1952), pp. 410-414.

perceived sets and boundaries in the physical world affect behavior as well.[45]

In exploring the social negotiations that lead to a sense of reality, authors such as Garfinkel have pointed out the extent to which "rationality" is a retrospective social reconstruction of what happened rather than some characteristic built in ahead of time.[46] Rules and principles are invoked afterwards to rationalize what we have done. If this is the case, it goes a long way toward explaining the frustration encountered in trying to plan organization structures according to the rational principles laid down by the early administrative theorists.

In this view also, people comprehend their experience largely through the construction of categories and relations among them; i.e., objects and events are typified rather than processed in all their detail. Such categorization serves the crucial function of simplifying the world for the thinker by reducing it to a manageable number of categories and relations to be processed and remembered.[47]

Although the propositions that can be generated from this perspective are not yet as fully researched as those in the other perspectives, they are worth looking into, for they suggest ways out of problems posed by the other perspectives.

Ranson argues, for example, that structural changes will follow changes in general interpretive schemes: "when priests revise their theology, when teachers adopt a more radical pedagogical frame of reference . . . we may expect structural forms to be altered to ensure their symbolic appropriateness." He also argues that structural changes will follow from "contradictory imperatives of situational constraints . . . for example, where an organization is large, and therefore constrained to become more bureaucratic, and at the same time is located in a turbulent environment and therefore constrained to become more flexible and adaptable in its structural arrangements."[48] The latter problem, as we noted earlier, definitely pertains to marine policy-making in the federal government. Meyer and Rowan hypothesize that in a highly institutionalized

---

[45] Robert H. Somer, *Personal Space: The Behavioral Basis of Design* (Englewood Cliffs, N.J.: Prentice-Hall, 1969); Donald Stone and Alice Stone, "The Administration of Chairs," *Public Administration Review* 34 (1974): 71.

[46] Harold Garfinkel is generally credited with being the father of a field called "ethnomethodology." See his *Studies in Ethnomethodology* (Englewood Cliffs, N.J.: Prentice-Hall, 1967). Further references may be found in Richard Harvey Brown, "Bureaucracy as Praxis" and in Stewart Ranson et al., "The Structuring of Organizational Structures."

[47] See J. Richard Eiser and Wolfgang Stroebe, *Categorization and Social Judgement* (London: Academic Press, 1972); Karl E. Weick, *The Social Psychology of Organizing*; Christopher Hood, *The Limits of Administration*; and Harold Schroder, Michael Driver, and Siegfried Streufert, *Human Information Processing* (New York: Holt, Rinehart, and Winston, 1967).

[48] Ranson, Hinings, and Greenwood, "The Structuring of Organization Structures." p. 12.

environment, such as state and federal governments and large university systems, structural units keyed to legitimacy in the environment may serve the organization's survival better than structural units keyed to workflow efficiency.[49]

I have argued elsewhere that one should focus on the alternative possible groupings of people, things, and activities in an organization and then ask which such groupings are most likely to survive, or emerge from, the social interaction processes.[50]

For example, it seems likely that social interaction will settle on boundaries that are perceptible in more than one way. There is a tendency to settle on differently based groupings whose boundaries coincide, because there are then multiple clues to the common identity of the circumscribed items. A colleague at another university reports that her political science department is split into two different groups, at opposite ends of the age spectrum, on opposite sides of the dispute between traditionalists and behavioralists, and located in different buildings. Given the coincidence of so many boundaries, a bifurcated view of the department could hardly be avoided by any member or observer. In a marine-related example, the difference between commercial and sport fishermen's groups is accentuated by the fact that it largely (though not entirely) coincides with the difference between saltwater and freshwater fishing.

As another proposition, if the items in some groupings are perceived to be strongly linked physically or socially, then that grouping is likely to survive in the sortings and reconstructions that go on. Such a set would be perceived as difficult or costly to split up. No one seriously suggests putting all the second basemen in the National League into one coherent group, because the disruption to the perceived working unit, the team, would be intolerable. The extent of interdependence among marine activities, as compared to their interdependence with land-based activities, is a central question, as we have noted before. The answer lies in personal perceptions and social negotiations.

Such factors as the legitimacy of the grouping and the familiarity of the group may also prove to be important. Meyer and Rowan argue, as noted, that large-scale "institutionalized" organizations have many units (such as affirmative action offices) that are erected more to give the organization legitimacy than to facilitate its work directly. The familiarity of the Coast

---

[49] John W. Meyer and Brian Rowan, "Institutionalized Organizations: Formal Structure as Myth and Ceremony," *American Journal of Sociology* 83 (1977): 240-363.

[50] Stuart A. Ross, "Organization Design in the California Energy Commission: The Adequacy of Alternative Design Perspectives," Ph.D. Dissertation, University of California, Berkeley, 1978.

Guard in the minds of the public and politicians is part of what makes it unlikely that in any marine reorganization the Coast Guard will be disbanded and its pieces sent to various other agencies.

Finally, it may be that organization members prefer consistent emotional or logical relations among the groupings they perceive and will resist proposed structures that require members to hold inconsistent relations in mind. Considerable evidence from social psychology suggests that people strive to maintain some kind of consistency among the relationships they perceive.[51] Situations involving inconsistencies are thought to be tension-generating, and the psychologists' argument is that people will take action to lessen or remove the tension by selective perception, by attitude changes, by terminating one of the relationships, or even by keeping the two relationships apart in time and space. Some of these actions may prove to be inappropriate for public policy.

As an example, can a government agency successfully carry out both regulation and promotion of the same constituency? The reorganization of the Atomic Energy Commission, in 1974, into separate regulatory and promotional agencies (the Nuclear Regulatory Commission and the Energy Research and Development Administration) was a prominent and recent example of concern with this question. In that case, the inconsistency was resolved by the apparent elimination of any affective link between the two new government agencies. Effective marine policy involves both regulatory and promotional tasks, often both with the same constituency, e.g., the offshore oil industry; how such affective inconsistencies are handled will be of great importance.

Thus, the approach proposed here provides several possibilities. For one thing, it integrates the other approaches by deemphasizing the difference between them, by suggesting that the content of the groupings we perceive (people, behavior, techniques, objects) is less important than the boundaries and other perceptual characteristics of those sets. Some workflow characteristics in some situations are important in shaping social consensus, and others are less so (if, for example, they are simply not perceived as important); some political factors in some situations are important in shaping consensus, and others are less so. Ranson deemphasizes, similarly, the oft-mentioned contrast between "formal" and "informal" structures. This perspective could explain why the propositions of each theory type sometimes meet with success and sometimes do not.[52] The factors emphasized by the other perspectives can

---

[51] For a good summary of this literature, see Marvin E. Shaw and Philip R. Costanzo, *Theories of Social Psychology* (New York: McGraw-Hill, 1970), especially chapter 8.

[52] See, for example, Edgar H. Schein, *Organization Psychology* (Englewood Cliffs, N.J.:

be and often are important to organizing, but they are not automatically special, central, or determining.

The approach has yet another advantage. In any organization, no one of the other theories can account for all of the units and structures—one finds a "functional" office that survived a "regional" reorganization, a cantankerous engineering group in the midst of an otherwise mellow organization, or a blatant patronage appointment in an otherwise rationalistic structure. Such seeming anomalies are understandable if they are seen as having the perceptual properties requisite for being selected in the social construction process.

Finally, the perspective lends theoretical respectability to the process of selling one's reorganization ideas—a process we see around us all the time. When a bureaucrat attacks someone else's design because it cuts across established dividing lines, when he defends his own design by reference to symbols of great legitimacy, he is taking us into the very guts of the reorganization process, not distracting us from them.

This perspective, more than any of the others, moves us away from the notion that only top managers or their advisers get to design organizations. It presents the contrary view that each organizational participant is an organization designer, in that his interactions help or hinder the emergence of particular structures. Designers vary in the scope of their designs, the resources they can bring to bear, and the extent to which their designs affect other persons in the organization, but they are all designers nonetheless.

There are, of course, important limitations to this perspective—limitations that make predictions and recommendations hard to derive.

First, the image of a social construction process leaves out the analysis of which persons have more to say than others in reaching the social agreement, which is simply the matter of power. If matters are such that in fact the top manager's views dominate the social processes, then discussing the breadth of the processes misses the point, ignoring matters that the organizational-politics perspective has more correctly identified. Ranson and Brown explicitly incorporate politics in their views but without much gain in precision.

Second, if one is to predict or recommend new structures by understanding the experiences and cognitive styles of the audiences for which they are intended, then rather onerous data requirements are involved. One is limited either to close-up case studies or to broad-brush treatments for larger groups. Neither is likely to be helpful for the marine reorganization problem as we have addressed it.

Prentice-Hall, 1970) or Robert H. Guest, *Organizational Change: The Effect of Successful Leadership* (Homewood, Illinois: Irwin-Dorsey, 1962).

Third, the theories as developed so far give little clue as to the relative importance of the various social-psychological factors, singly or in combination. Do changes in ideology always lead to changes in structure? Is consistency more important than familiarity, and for which audiences, and to what extent?

Perhaps further empirical research in this perspective will resolve some of the difficulties. But until the difficulties are resolved, this approach may also be nonfalsifiable in nature and, hence, not very helpful.

Final Comments

Five ways of thinking about reorganization have been presented: reorganization as a rational exercise in achieving efficiency, as a political result, as a support for human needs, as a response to technological or environmental constraints, and as a continuing social process. Each way captures some but not all of the reorganization problem, and each overlaps with the others. Believing one of them while ignoring the others will surely create difficulties and failures that could have been avoided. So, too, will merging the approaches thoughtlessly. No universally successful path to reorganization has been presented because none has been found by anyone. The opportunities and obstacles in the path to reorganization—the successes and failures of others—have been indicated, however, and that knowledge can at least make marine organization efforts more sophisticated, which may be achievement enough.

CHAPTER 5
AN ECOLOGY OF GOVERNMENTS: COASTAL ZONE MANAGEMENT IN A FEDERAL SYSTEM

ROBERT WARREN*

Introduction

A number of major policy issues facing the United States are marine-related and, in spatial terms, many of them are located within the nation's coastal zone. Energy facilities siting, competing recreational, commercial, and residential urban land and water uses, and environmental protection are but a few of the issues which now are commonly recognized as leading problems in coastal areas.

The inclusion of these problems on the agenda of national concerns is, however, relatively recent. Widespread awareness of conflicts over biological, economic, and social values growing out of the use of coastal resources coincided with a rapid increase in their number and magnitude and with the rise of a citizen-based environmental movement in the late 1960s. One initial response was to seek an institutional means for resolving the conflicts. Thus, coastal zone management was "invented" a little more than a decade ago.[1] The coining of the phrase, however, preceded general agreement on its content and the desirable organizational arrangements for its application.

There is, on the one hand, wide agreement that coastal zone resources are of great value and that the public interest must be reflected in their allocation through governmental regulation. Yet, on the other hand, there is no consensus on what mix of uses and users is most appropriate or on the form, type, and scale of governmental involvement that should be utilized to make decisions about and to administer coastal policies. Consequently, efforts to design coastal management systems have

* Robert Warren is Professor of Urban Affairs and Marine Studies at the University of Delaware. An earlier version of this paper was prepared for the conference Formulating Marine Policy: Limitations to Rational Decision Making, sponsored by the Center for Ocean Management Studies, University of Rhode Island, in cooperation with the Office of Ocean Management and Office of Sea Grant of the National Oceanic and Atmospheric Administration, Narragansett Bay Campus, University of Rhode Island, Narragansett, Rhode Island, June 19-21, 1978.

[1] One of the initial summaries of coastal management problems is contained in Bostwick H. Ketchum, ed., *The Water's Edge: Critical Problems of the Coastal Zone* (Cambridge, Mass.: MIT Press, 1972).

required experimentation with governmental organization and resource use criteria.

From a governmental perspective, the choice of mechanisms to regulate activities in the coastal zone deals with basic questions about the distribution of authority among the local, state, and national levels in our federal system. Curiously, these issues have received far less attention than is warranted. Coastal land and water run through the boundaries of innumerable cities and counties as well as numerous states. If resource use within the coastal zone is to be determined on the basis of national, state, or supra-local regional decisions, some redistribution of authority within the federal system must take place.

Transferring some or all of the control to a larger scale of government—the most frequent proposal made—is more than a management issue. It encompasses broader governance questions. In fact, we may be close to significantly modifying important aspects of the balance among local, state, and national decision-making authority by shifting control upward without ever having seriously debated or, perhaps, without even realizing the nature of the values involved.

In the early 1970s three states—California, Delaware, and Washington—on their own initiative, created quite different statewide management structures for regulating land and water uses in their coastal areas.[2] Federal action was being considered at the same time, and the Coastal Zone Management Act of 1972 (CZMA) resulted. It placed a national priority on protecting the natural, biological, and physical resources of the coast. Much of the control over these phenomena, however, rests with state and local governments under the constitution. Consequently, national coastal zone policies and guidelines for resource use were designed to be carried out voluntarily by state governments in exchange for fiscal and other incentives. Provisions of the act have been attractive enough to induce all thirty eligible states to engage in some type of effort to establish a coastal zone management system.[3] Section 305 of the act provides grants to states of up to 80 percent of the costs for the development of their programs. Once they are approved by the

[2] Robert L. Bish, Robert Warren, Louis F. Weschler, James A. Crutchfield, and Peter Harrison, *Coastal Resource Use* (Seattle: University of Washington Press, 1975), chapter 8; Stanley Scott, *Governing California's Coast* (Berkeley: Institute of Governmental Studies, University of California, 1975), chapter 1; and *Coastal Zone Act Administration, June 28, 1971–June 30, 1977* (State Coastal Zone Control Board and Office of Management, Budget and Planning, Delaware, November 1977).

[3] By early 1980, 16 of the 30 states had OCZM approval for their management programs and 10 were being processed. The remaining four had either withdrawn or were involved in protracted conflict over the acceptability of their proposals. See Sarah Chasis, "The Coastal Zone Management Act," *Journal of the American Planning Association* 46(1980):147.

federal Office of Coastal Zone Management, Section 306 makes available funds for their administration. Another of the incentives for state action in the CZMA is Section 307, the "federal consistency" requirement. For approval by the OCZM, a state's management plan must indicate that priority will be given to matters of "national interest" in decisions on coastal resource utilization. In return, federal agencies are to become subject to state regulation under the plan.

These efforts to bring state policies in line with nationally determined priorities also have third-party governmental effects. In order to satisfy CZMA requirements, states are expected to enter into land use policy and regulate activities which have traditionally been controlled by local governments. The means of broadening federal influence over coastal development are a series of provisions in the act. Among other things, it mandates that states assure that local policies will not unreasonably restrict coastal uses for regional benefit [Section 306(e)(2)] and that decisions on the siting of energy-related and other facilities in coastal areas will give adequate consideration to national over local interests [Section 306 (c) (8)]. To be approved by the OCZM, a state's program must show that it possesses the authority to control coastal development and to obtain local compliance with its provisions.

Underlying the design of the CZMA and coastal management programs initiated within states is a belief that public problems are best solved by transferring decision-making powers from smaller to larger scale organizations. The resulting proposals for structural change, however, have been based on a number of uncritically accepted assumptions concerning: (1) the benefits of large scale organization for policy formulation and administration; (2) our capacity to simplify complex governing systems without producing negative effects; and (3) that federal actions necessarily reflect the national interest. A closer look at the recent efforts to modify the distribution of governmental authority for allocating coastal resources will allow an elaboration of these points.

Governing the Coastal Zone

Decisions concerning the use of coastal resources have been and still are basically made by a complex ecology of governments which vary considerably in terms of scale, organizational type, and mission. Coastal cities and counties exercise considerable control over their shorelands through land use authority, police powers, and the public goods and services they choose to produce. Independent authorities, such as port districts, may exist for limited but important functional purposes. State and federal agencies produce relevant goods and services, such as water-oriented parks and recreation, harbor improvements, and marine rescue services. The major influence of state and federal policies in the coastal zone, however, is through the regulation of the activities of others in both the public and private sectors and providing incentives for specific actions by state and local governmental units with grants and other types of funding.

In the Los Angeles area, for example, in addition to numerous cities and the county, the following governmental units exercise authority over portions of the coastal zone: the South Coast Regional Commission and State Coastal Commission; the Southern California Association of Governments; the Metropolitan Water District; a county flood control district; the South Coastal Air Quality District; a regional water quality control board; and the California Energy Commission; not to mention the state legislature.[4] Among relevant federal agencies are the Bureau of Land Management, the Coast Guard, the Environmental Protection Agency, the Fish and Wildlife Service, the Department of Housing and Urban Development, the Corps of Engineers, the Maritime Administration, the Department of Defense, the Heritage Conservation and Recreation Service, the Federal Energy Regulatory Commission, the Nuclear Regulatory Commission, and the Office of Coastal Zone Management.[5]

Even this lengthy list does not adequately describe the actual complexity of the coastal governing process which involves the interaction of these units with each other and the broader political and judicial systems within which they are embedded.[6] Administrative and regulatory agencies ultimately are responsible to elected officials and potentially are susceptible to the influence of interest groups. Further, bargaining among public agencies is an important means of resolving policy conflict. At the same time, many direct allocative decisions are made through market transaction in the private sector. The pricing system plays a major part in determining what use will be made of particular parts of the coastal zone and who will have access to the market. Finally, adjudication in state and federal courts often has a dominant role in creating policies and making

---

[4] Louis F. Weschler, "Environmental Quality Within an Intergovernmental Matrix: A Case Study of the SOHIO Project," *Coastal Zone Management Journal* 6 (1979):236,237.

[5] Ann Breen Cowey, "The Urban Coast from a National Perspective," *Coastal Zone Management Journal* 6 (1979):135-165.

[6] For a case study of this complex interaction, see: Weschler, "Environmental Quality," pp. 233-252.

specific allocative decisions which are binding upon public agencies as well as private citizens and firms.[7]

The output of this complex system is, in the minds of coastal zone management proponents, unacceptable because it does not give adequate weight to environmental values. This failure in the public sector is attributed to the existence of too many governmental units with control over segments of the coastal zone. Some, such as local governments, are seen as being organized on too small a scale and as being too responsive to political pressures which result in decisions that do not take environmental values into account. Even at the state and federal levels, the argument goes, there is an excess of agencies dealing with coastal matters that have overlapping and sometimes conflicting responsibilities. The centralization of control is again called for, this time to place clear-cut management powers in a single agency. Two of the key elements in proposals by coastal zone management advocates relating to governmental structure then, have been the transfer of control to larger scale organizations to better reflect regional, state, or national environmental interests and concentrating as much authority as possible in a single agency.

These policy recommendations contain several ironies. One is that groups concerned with energy production and supply are making similar arguments for state, if not federal, control over all energy facilities siting decisions in the coastal zone. This strategy has the opposite motivation. Energy advocates believe that the existing system with multiple decision points at the state and local levels allow these units, especially localities, to place environmental values above energy supply priorities in their policies toward coastal land use. The energy-oriented model of policy making assumes that environmental values will be more predictably subordinated to energy production and undue delays and costs now imposed upon energy firms will be eliminated if control over facility siting is as centralized as possible at the state or national levels. This issue will be considered in more detail below.

The second irony relates to the contrasting assumptions that environmentalists tend to have about human and non-human ecological systems. Exactly the opposite model is adopted for marine ecological systems as for the management and policy systems that are called for to protect the former. The actions of people are believed to be endangering

---

[7] A more detailed discussion of these decision processes that effect the allocation of coastal resources is contained in Robert Warren, Robert L. Bish, Lyle E. Craine, and Mitchell L. Moss, "Allocating Coastal Resources: Trade-Off and Rationing Processes," in Ketchum, ed., *The Water's Edge*, pp. 233-240.

marine ecosystems by making them more fragile, unstable, and vulnerable through eliminating components, interfering with processes, and simplifying their structure. A biologist will seldom be willing to support actions that would substantially change the habitat of a major species of marine life without some understanding of how the ecosystem will be affected. Proposals to change and simplify the structure and processes by which we make and administer policies for coastal resource utilization should be treated with a similar caution.

In human social, economic, and political systems, productivity, cultural achievement, and the generation and utilization of knowledge all tend to be related to systemic complexity.[8] This suggests that action to reduce the complexity of public sector systems should be made only with an understanding of how the present allocative processes function and with the aim of preserving the beneficial aspects of complexity. Two examples can be used to pursue this analogy. One relates to the initial experience with a state coastal zone management program in Los Angeles County. The other concerns the role of local and state governments in the siting of energy facilities in the coastal zone.

Centralizing Urban Coastal Decision-Making

California has been in the forefront of both coastal resource exploitation and regulation. In 1972 the voters of the state enacted the most comprehensive and centralized coastal management system yet adopted. It transferred final authority over land extending a thousand yards inland from the mean high tide mark from cities and counties to regional and state-level agencies. In the referendum campaign for Proposition 20, two themes were heavily emphasized. One focused on the claim that great amounts of rural and underdeveloped, as well as urban, shoreline were being transformed to uses that were unacceptable in ecological and aesthetic terms or that reduced public access. The second theme dealt with the causes of this misallocation of coastal resources. Decisions by cities and counties about coastal development were seen as a primary source of the problems. A basic premise of the proposition was that local governments must give way to decision-makers who have a broader perspective and larger scale boundaries.

In the logic of this view, implicitly at least, local governments constitute a homogenous set. What can be said about one can be said

---

[8] John D. McEwan, "The Cybernetics of Self-Organizing Systems," in C. George Benello and Dimitrious Roussopoulos, eds., *The Case for Participatory Democracy, Economic and Social Development* (Baltimore: Johns Hopkins Press, 1971).

about any other in the sense that all city and county officials are seen as susceptible to market pressures for the development of coastal land to its highest economic use. Alternatively, if communities respond to locally determined noneconomic priorities, they tend to be narrow and in conflict with the broader interests of the region or state.

This "breakdown model" of the public sector at the local level has been central to the rhetoric of supporters of statewide coastal management systems, not only in California but throughout the country.[9] One of the effects of this perspective has been to orient the design of coastal regulation to the protection of nonurban areas and to make little differentiation among local governments in terms of their impacts upon coastal resource and the use patterns of urban residents.

The 1972 legislation paid little attention to the question of how to meet the socioeconomic and environmental preference of subsets of the population in urban areas of one, two, five, or ten million people and reconcile them with each other and state or national interests. This law, which literally removed final control over their shorelines from cities and counties, was enacted with little knowledge of the behavior of local governments in large urban centers or how they varied from one another and from those in less developed portions of the state.

Evidence of the gap between the assumptions in the breakdown model upon which Proposition 20 was based and the actual characteristics of the allocative system for coastal resources in a metropolitan area is reflected in a study of local governments in the Los Angeles area.[10] The research sought to test an alternative set of assumptions. In reference to Los Angeles County, it was postulated that city and county governments have the capacity and, in fact, frequently do consciously adopt values concerning their shorelines that mediate market forces and result in heterogeneous sets of uses that clearly distinguish one portion of the metropolitan area from another. The political processes of local governments were viewed as mechanisms which allowed diverse preferences within and among communities in the same region to be translated into discrete combinations of public and private goods and services in the coastal zone. Further, it was hypothesized that legislation which seeks to simplify the policy formulation structure of a region without fully understanding the functions of the preexisting allocative system can have effects on communities that are unanticipated, not uniform and dysfunctional for the community and region as a whole. The

[9] Robert Warren, Louis F. Weschler, and Mark S. Rosentraub, "Local-Regional Interaction in the Development of Coast Land Use Policies: A Case Study of Metropolitan Los Angeles," *Coastal Zone Management Journal* 3(4):352,353,356,358.
[10] Ibid., 353-379.

study utilized records of development permit action by the South Coast Regional Commission between 1973 and 1975 as well as historical data on the ways in which nine cities and the county government typically allowed their coastlines to develop prior to the new law.

The results offered little support for the breakdown model as a description of the performance of local governments in the region. An exceedingly complex and spatially differentiated set of uses existed along the metropolitan coast in 1973 when Proposition 20 was implemented. They could not be explained as the outcomes of unregulated market forces or community decision-makers seeking only short-term, narrowly defined local benefits. Rather, they reflect the interplay among market pressures, regional demands for coastal-related public goods and services, and community preferences articulated through local political processes. The results ranged from well-defined positions for or against intensive or extensive development to total indifference. Some policies had remained stable for decades, others had changed at various times or were in transition.

An analysis of the permits issued by the regional commission indicated that few marked changes occurred in previously established mixes of housing, commercial, and industrial activity within communities. Yet, the political and economic costs required of each city and the county were far from equal in maintaining their preferred coastal resource utilization patterns. They varied with the amount of each community's land area and resources that were within the regulated zone, the type of policy followed locally, the extent to which the subarea's image was associated with its coastal location, and its ability to wield political power in the region.

The law represented an attempt to replace one allocative system with another without full comprehension of either what was being foregone or the equity issues that would arise in relation to its differing impacts. The recognition of the importance of local governments in articulating subregional preferences and evaluating the effects of alternative use patterns does not mean that all locally preferred policies must prevail. Rather, it indicates that coastal policy formulation in highly urbanized areas may be a problem of resolving differences and establishing cooperative interaction among *competent* publics of differing scale—local, regional, and state—as opposed to one of simply overcoming opposition to the transfer of control upward.

One unique aspect of the 1972 California law was its self-liquidating provision, specifying that it should not extend beyond five years. As a result, the position of local government was substantially strengthened in the California Coastal Act of 1976. This occurred, however, only after intense conflict in the legislature over the role of local government. Cities are now responsible for developing local coastal plans and the issuance of coastal development permits, subject to criteria and guidelines established by the legislature and a statewide commission.[11]

Energy Siting Policy and Local Government

The desirability of adopting policies that will increase the supply of energy available in the United States and reduce or eliminate dependence on external suppliers has few dissenters. This unanimity does not extend, however, to specifics such as which energy sources should be increased and by what means. Many of the resultant conflicts have dealt with the matter where energy facilities should be sited and who should make such decisions. In fact, some of the most visible and volatile questions concerning coastal development have grown out of the disputes over whether receiving, processing, storage, distribution, and support facilities for oil, gas, and nuclear energy should be located in the coastal zone and, if so, where.

Proponents of a rapid expansion of energy supply have commonly backed proposals for shifting control over energy-related policies from smaller to larger scales of government. Their rationale parallels that offered in favor of similar upward transfers of power for coastal management systems. It is argued that a reduction in both the number of governmental units participating in decisions and the complexity of the permit process will enhance efficiency, equity, and the national interest. Local governments, again, are clearly the odd person out. Localities would have to give way to states and, if necessary, the states to the federal government.

The same questions raised about the implications for the federal system of the redistribution of authority associated with coastal management are relevant here. This is particularly evident in terms of the way national policies have been pursued in the development of Outer Continental Shelf (OCS) oil and gas resources. It also is a case in which coastal management and energy supply values are in conflict.

Imperial Federalism

A national policy decision was made in the mid-1970s to place a priority on the development of oil and gas resources presumed to exist on the OCS. The plan involved the acceleration of the leasing of a substantial

---

[11] Stanley Scott, "Notes on California's Coastal Governance: A Reply to Peter Douglas," *Coastal Zone Management Journal*, 4(4):475-486.

number of sites in many coastal sections of the country to energy firms by competitive bidding. This would be followed by exploration and, hopefully, production stages. In the latter two, onshore support systems would be required.

The massive OCS exploration and production that was envisioned resulted in considerable controversy over how potentially negative onshore, as well as offshore, environmental and socioeconomic impacts could be reduced or avoided.[12] Part of the conflict dealt with who would control the process. This dispute was framed exclusively in terms of a state-federal contest over authority to regulate OCS development. The matter was settled in *United States v. Maine* by the Supreme Court in 1975. Federal jurisdiction over decisions on OCS site leasing and offshore exploration and production was upheld.

Treating the debate over control of OCS development as a federal versus state issue put local governments in an anomalous position in two senses. On the one hand, they have no standing in relation to OCS development decisions even though they will be directly affected by them. The location, type, and magnitude of offshore exploration and production will require onshore support systems that will impact coastal communities. The federal government has done nothing to provide a formal role for localities in the shaping of its offshore policies. For that matter, some federal agencies are not reconciled to a consultative status for state governments prior to leasing decisions even though there is a legal basis for it.[13] On the other hand, federal officials have sought to formulate OCS-related onshore, as well as offshore, policies as if localities possessed no relevant authority. This is not the case.

State and local governments have constitutionally derived powers that give them primary control over the creation and operation of onshore support systems. This is true of local governments through their land use and zoning powers and through their authority to produce municipal goods and services. Yet, national policy makers have treated cities and counties as a residual category in the federal system for energy decision making and planning. To do so they have had to define the role and probable behavior of localities on energy-related matters in ways that deviate considerably from reality.

Federal officials have assumed that communities are amenable to externally defined priorities and are capable and willing (or can be made

---

[12] Pamela L. and Malcolm F. Baldwin, *Onshore Planning for Offshore Oil: Lessons from Scotland* (Washington, D.C.: The Conservation Foundation, 1975); and James K. Mitchell, "Onshore Impacts of Scottish Offshore Oil: Planning Implications for the Middle Atlantic States," *Journal of the American Institute of Planners* 42(1976):386-398.

[13] Chasis, "The Coast Zone Management Act," p. 152.

so) to establish and manage the onshore infrastructure needed for energy supply expansion. In effect, they would accept facility siting decisions determined by the preferences of national agencies and private sector firms. Underlying this set of assumptions is an attitude toward subnational governments that can best be described as "imperial federalism." It is based on the belief that when an understanding or agreement is reached by decision makers at a higher level in the federal system it can be ascribed to all lower level components. This mindset carries with it the view that things can be done *to* or *for* rather than *with* local communities to satisfy federal mandates in energy policy. However, this perspective has had limited predictive capacity. The behavior of subnational units toward energy siting proposals has produced a number of "surprises" for both national agencies and energy firms that extend well beyond OCS-related cases.

Obtaining permits from the Bureau of Land Management (BLM) for offshore drilling does not, for instance, automatically guarantee state approval of onshore support facilities. The conflict between the California Coastal Conservation Commission and Exxon in 1976 over the specifications for onshore pipelines to support BLM approved oil and gas production off Santa Barbara is a case in point.[14] Neither does strong political backing from state officials for the construction of a coastal refinery ensure acceptance of it by a community. The widely celebrated referendum vote in Durham, New Hampshire, that prevented Aristotle Onassis' Olympia Oil Refineries, Inc., from locating a facility there, in spite of wide support from state leaders, is only one of a number of such incidents.[15] A review of ten refineries that were planned but not constructed on the East Coast shows that one-half of these projects were blocked by local government or community action groups in the early 1970s.[16] Four more proposed energy facilities were rejected because of local opposition in Hudson County, New Jersey, alone between 1974 and 1976.[17]

[14] Robert G. Healy, "An Economic Interpretation of the California Coastal Commissions," in Robert G. Healy, ed., *Protecting the Golden Shore* (Washington, D.C.: The Conservation Foundation, 1978), pp. 146-148.

[15] An article in the July 1974 issue of *Esquire* magazine by Gerry Nadel is entitled "The Score from New Hampshire: Democracy 1, Aristotle Onassis 0."

[16] Robert B. Biggs, "Offshore Industrial-Port Islands," *Oceanus*, 19(1975):60.

[17] David Morell and Grace Singer, *Alternative Energy Facility Siting Policies for Urban Coastal Areas: An Executive Summary of Findings and Policy Recommendations*, Princeton, N.J., Center for Environmental Studies, Princeton University, PU/CES 83, March 1979, p. 7.

Negative Feedback and the Federal Response

The resistance of local officials and community groups to the location of energy facilities can be interpreted and responded to in several ways. One way is to view it as a few communities capriciously placing local above national interests. The solution, then, is not to modify federal policy toward localities but to increase the incentives for local conformity or seek to eliminate their control over energy-related land use decisions. These have been the characteristic responses at the national level. An alternative interpretation of the above phenomena is that community opposition has been widespread and recurring enough to constitute an important unresolved problem in our governing process. This perspective suggests a different explanation of community behavior and a reevaluation of federal policies.

Local communities have legitimate reason to be concerned with the position of those backing swift expansion of energy production that localities *must* accept facilities and, perhaps, receive compensation for some portion of the resulting negative spillovers. Apprehension is appropriate also when federal administrators who devise facility location policies ignore the existing legal authority of cities and counties in the matter. This concern of local leaders has frequently been translated into strategic behavior to protect community values.

Localities have adopted land use regulations that explicitly prohibit energy-related facilities. In other instances, where referendum provisions are available, proposals have been vetoed by local voters. Cities and counties also can create a negative climate by making it clear to energy firms that they are not welcome. Even if this fails to deter efforts to establish a facility and decision making authority rests elsewhere, intense local opposition can decisively influence the actions of higher level authorities.

The courts also are frequently used by communities and groups seeking to prevent the siting of an energy facility.[18] A more general set of issues was raised by Suffolk County, New York, when it sought and obtained an injunction in early 1977 to prevent OCS lease Sale No. 40 in

---

[18] In 1978 it was estimated that 300 court suits had been brought by states, localities and environmental groups claiming that the Environmental Impact Statements required of federal projects by the National Environmental Policy Act were inadequate. Approximately 400 suits have argued that the mandated Environmental Impact Statements had not been completed on other projects and sought compliance. While not all of these involved coastal area projects, a number were located in the coastal zone. See James K. Mitchell, "Impact of Offshore Oil and Gas Development on the Coastal Zone: Reforming the Impact Assessment Process," *Coastal Zone Management Journal* 4(3): 309.

the mid-Atlantic area.[19] The county argued, in part, that there was not appropriate consideration in the BLM Environmental Impact Statement of the authority and possible response of local governments to onshore support system needs for the exploration and production stages of Site No. 40-related developments. While the injunction was reversed later in the year, the action by the county did force, for the first time, the discussion of a number of issues concerning the role and interests of local governments in OCS development.[20]

This is not to say that all communities oppose all energy facilities or that they do so successfully. Many are bidding eagerly for them.[21] Rather, it reflects the fact that the image of national policy makers of local governments as tractable has frequently proven wrong. More serious, however, the response of federal officials has not been to conclude that too little knowledge exists about local behavior, preferences and on-site information about facility impacts or that too little attention has been given to decision and planning procedures that can adequately reflect legitimate local interests. Typically, the response has been to assume that local governments can be made supportive or, at least, neutral toward energy facilities by appearing to shower federally generated data and compensatory funds upon communities. The promise, however, has been far greater than the performance in each case. The numerous federal projections of the expected socioeconomic and environmental impacts of energy facilities from OCS development upon local communities have seldom proven usable by the localities themselves.[22] The Coastal Energy Impact Funds authorized in 1976 amendments to the CZMA to enable states and their affected localities to mitigate costs resulting from the siting or expansion of energy facilities have provided little benefit to most states, and the status of local governments in the distribution of such funds has never been adequately defined.[23] Money for states to plan and operate their coastal management programs and, by definition to constrain local autonomy, has been much more predictably available to all participating states.

A good deal has been made of the federal consistency provision of the CZMA as a carrot to gain the agreement of states to recognize the

[19] *County of Suffolk v. Secretary of the Interior*, U.S. District Court, Eastern District of New York, No. 5, 75-C-208, 75-C-229 (February 17, 1977).

[20] County of Suffolk v Secretary of the Interior, U.S. Court of Appeals, Second Circuit, Nos. 77-6049, 77-6065 (August 25, 1977).

[21] Communities on the Texas-Louisiana Gulf Coast, for example, have been much more favorable to energy facilities than those on the East Coast. See Morell and Singer, *Alternative Energy Facility Siting*, pp. 6-13.

[22] Mitchell, "Impact of Offshore Oil," pp. 313-323.

[23] Chasis, "The Coastal Zone Management Act," p. 152.

national interest in their coastal policies. This provision clearly implies that the states will be expected to support, if not impose, facility siting decisions favored in federal policy. At the same time, the consistency requirement should provide states with greater general control over the activities of federal agencies in their coastal areas. Federal commitments, however, have been almost exclusively weighted toward the former rather than the latter part of the exchange. The benefits for state from the federal consistency clause thus far have proven illusory.[24]

The Scale of Decision-Making and the National Interest

It is clear that many communities and some states are opposed to or will critically evaluate proposals for new energy facilities. The action may be based on economic, environmental, or community revitalization goals, or on safety factors, among others. Such behavior hardly satisfies the national interest if that interest is defined exclusively in terms of increased energy supply. Further, to adopt this criterion would require significant modification in the distribution of power within the federal system.

The federal response to local opposition to energy facilities is analogous to killing the bearer of bad news rather than dealing with the problems of which the unfortunate messenger brings reports. Large scale, hierarchically structured organizations, such as the federal agencies involved in OCS development are notorious for their capacity to distort information as it moves up channels and to reject negative feedback from the environment.[25] Eliminating the ability of localities to veto externally made siting decisions and failing to provide them with any participation in the new decision process will seriously reduce the probability that all costs imposed upon the people and communities in the immediate area will be identified and compensated for or ameliorated.

It is dangerous to assume that the regional, state, or national interest in coastal-related decisions will always be better reflected by transferring authority from smaller to larger scale jurisdictions. This is particularly true if no attention is given to establishing means by which the preferences and on-site knowledge of the smaller systems can be articulated and given standing in the enlarged decision process.

---

[24] Ibid.
[25] J. A. Stockfisch, *Incentives and Information Quality in Defense Management*, Santa Monica, Rand, R-1827-ARPA, August 1976; and Robert A. Rosenthal and Robert S. Weiss, "Problems of Organizational Feedback Processes," in Raymond A. Bauer, ed., *Social Indicators* (Cambridge: MIT Press, 1966), pp. 302-340.

The distribution of governing authority among national, state, and local levels involves a unique balancing of interests and powers. Attempts to deliberately eliminate the complexity and redundancy produced by multiple decision points can have costs for the performance of the overall system that are greater than the benefits. Such efforts also can result in unintended outcome. The removal of local communities from either coastal or energy decisions would not end conflict, delays, and stalemates. Conflicts would be transferred into legislative and judicial contests and direct bargaining among large scale citizen and environmental groups, federal agencies and energy firms.

Two concurrent problem areas—coastal resource management and energy supply—have resulted in strong and somewhat successful pressure for the redistribution of authority among the three tiers of the federal system. However, two quite separate subsystems have emerged. One, coastal management, is designed to deal as comprehensively as possible with a narrow band of land and water. The other, the energy supply decision system, has no such spatial restriction and overlaps the coastal zone. This has resulted in conflicts about which subsystem has jurisdiction over energy siting in coastal areas. The most common outcome has been for energy-related agencies to be given primary control, thus producing the type of fragmentation that the coastal management programs were intended to reduce.[26]

Governmental Impact Statements

Returning to the parallel between marine ecosystems and political systems, perhaps the supporters of proposals that would significantly modify the existing structure of government and distribution of authority within the federal system should be required to file a Governmental Impact Statement (GIS). The GIS would analyze, for example, the effects of regional coastal management agencies or the transfer of control over energy siting decisions upon the capacity of cities and counties to function as units of general government which are responsible to serve as mechanisms through which citizens can exercise control over the state of affairs of their community. If the impact is negative, how will necessary or desirable aspects of the governing process at the affected level be maintained? For instance, in the case of energy siting powers, how would the interests of the community most directly affected be represented in the larger decision system? Will local officials have direct membership on the external decision making body? A formal advisory role? Or will they

[26] Weschler, "Environmental Quality," pp. 243-251.

constitute one interest group among many lobbying with a regional, state or national agency over what is to be located within their boundaries? Who will have standing to negotiate with energy firms over impact mitigation policies for the development and operation of the facility?

The GIS should consider another perspective as well, by critically looking at the probability that the benefits which the proposed institutional changes are supposed to provide will be achieved. If the centralization of all decisions concerning coastal resources into a single agency is one of the important justifications for a coastal management program, we should know the likelihood that this will be gained by the changes that are being proposed. Realistically, either drastic modifications would have to be made in government to fully achieve this end or the goal of a single controlling agency will have to be reconsidered. If the latter is the case, will it affect the rationale for other aspects of the organizational design?

There already have been a number of identifiable shifts in authority away from the local level in the pursuit of coastal management and energy supply goals. The extent has varied in state coastal management programs. Few have gone as far as California, but the states are clearly serving as mechanisms for imposing federal priorities on themselves and their localities. The position of cities and counties is undergoing much more obvious erosion in the case of energy.

In 1972 a federal court ruled that a local authority (the Hackensack Meadowlands Development Commission, New Jersey) could not use its zoning and building permit authority to prevent a private utility with a federal power commission certificate of approval from building a liquified natural gas storage facility. Based on the *Transcontinental Gas v. Development Commission* (No. 71-2034) case decision a precedent exists for overruling local and even state zoning ordinances when they are found to interfere with the interstate transportation of energy-related products.[27] A 1977 study of state energy siting policies in the Northeast found a movement toward "one-stop" procedures for obtaining permits for energy facilities, with the state government able to override local vetoes of proposed locations.[28]

Looking at these trends in terms of California's experience, one observer has offered a scenario of the siting process in 1990 if these

[27] David Morell, *Who's in Charge? Governmental Capabilities to Make Energy Siting Decisions in New Jersey* (Princeton, N.J.: Center for Environmental Studies, Princeton University, PU/CES 48, July 1977), p. 63.

[28] David Morell and Grace Singer, *State Legislature and Energy Policy in the Northeast: Energy Facility Siting and Legislative Action*, (Upton, N.Y., Policy Analysis Division, National Center for Analysis of Energy Systems, Brookhaven National Laboratory, June 1977), pp. 246-247.

tendencies continue.[29] To build an energy-related facility in the coastal zone, a corporation would apply to the U.S. Department of Energy (DOE) for a Certificate of National Interest. The DOE Office of National Interest Calculations would evaluate the project in terms of a series of state, interstate, national, and worldwide computer models. If the printouts showed a favorable balance, a National Interest Determination Commission would issue a Certificate of National Interest for the project within a week of receipt of the data. The corporation then would apply to the National Industrial Facilities Site Bank for authorization to locate in the area of its preference. The Site Bank would select the most appropriate tract in the area on the basis of data from a national industrial site ranking project and authorize construction. At the same time, the bank would provide construction specifications on how to build the facility to meet all environmental regulations. The scenario concludes, noting that the corporation has only three steps in its shopping trip for full approval and the process takes only nine days. All other agencies and levels of government have long ago been preempted.

It should be made clear that the logic of viewing the governing processes for the allocation of coastal resources in an ecological framework is not to argue that the federal system is or can be made static. National, state, and local roles and powers will and should change over time. Rather, the point is that if we are to make proposals affecting coastal-related policies which will significantly modify the existing distribution of governmental authority, we must be aware of it and understand what is being foregone and how we wish to compensate for it. Similarly, we must have a better knowledge of the probability of success of such proposals for both gaining the expected benefits and ameliorating the anticipated costs.

[29] William R. Ahern, "Energy Facilities and the California Coastal Act" (Paper delivered at the Conference on the Urban Coast and Energy Alternatives, Princeton University, Princeton, N.J., May 1978), pp. 18, 19.

CHAPTER 6
THE DECISION-MAKING PROCESS AND THE FORMULATION OF MARINE POLICIES

GARRY D. BREWER*

For anyone familiar with the history and particular character of marine policy in the United States,[1] it would be hard to deny or quibble with the summary denunciation by the United States Comptroller General, who stated in 1975 that the United States "had no comprehensive national ocean program."[2] Rather, what has been consistently the case is a highly fragmented and politicized decision process that appears to be incapable of measuring up to the large and expanding set of problems that press for attention and resolution.

In the following admittedly brief discussion, elements of a general decision-making process are presented and are then set off against what appear to be the main institutional features of marine policy-making and execution.

Decision-Making as a Process

It is helpful to conceive of problems as having a "life cycle" during which they emerge, are defined and estimated in terms of the potential solutions, are confronted with strategic statements (policies) and tactical measures (programs) that are expected to reduce or resolve unwanted consequences, and, in time, end, stabilize, or worsen as a result of both corrective acts and changes in the problem setting itself.[3]

* Garry D. Brewer is Professor of Organization and Management, and Political Science at Yale University. This chapter was originally prepared as a presentation to a conference on "Formulating Marine Policy: Limitations to Rational Decision-Making," Center for Ocean Management Studies, University of Rhode Island, June 19-21, 1978. The conference was sponsored by the Center and the Office of Sea Grant of the National Oceanic and Atmospheric Administration. Permission to publish this revision is gratefully acknowledged.

[1] I have relied on a well-known if somewhat unconventional source for my preliminary musing on this general topic: M. S. McDougal and W. T. Burke, *The Public Order of the Oceans* (New Haven: Yale University Press, 1962).

[2] United States General Accounting Office, *The Need for a National Ocean Program and Plan* (Washington, D.C.: Government Printing Office, Report B-145099, October 1975).

[3] This "life cycle" concept is described in Garry D. Brewer, "The Policy Sciences Emerge: To Nurture and Structure a Discipline," *Policy Sciences*, 5 (September 1974): 239-244, where it is illustrated with problems from several distinct policy arenas.

Invention or Initiation, the earliest phase of the process, begins when a problem is first sensed. At this point, a number of ways to alleviate it may be proposed, including many ill-defined and inappropriate solutions. This phase, marked by a casting about for raw information and even more unrefined answers, should help to sharpen and refine the definition of the problem.

Estimation, the second logical step in the process, deals with risks, costs, and benefits associated with each solution suggested in the invention phase. Estimation implies narrowing the range of plausible solutions (by excluding the infeasible or the truly exploitative, for instance) and ordering the remaining options according to scientific *and* evaluative criteria. A battery of sophisticated methodologies, in varying states of development and of varying degrees of suitability or appropriateness, is available for these tasks.

Selection, the third phase, is most easily seen as the "political" step. Someone—usually the policy-maker or decision-maker—must select from the "invented" and "estimated" options. This individual (or group) must strike a balance between the analyst's rational calculations and the multiple, changing, and conflicting goals of those having a stake in the problem and in society at large.[4] *Implementation* refers to the execution of the selected option. As evidenced by heightened interest in and statements about the failures of policy implementation, this is a phase of the overall policy of decision process that is little understood, not particularly appreciated, and not well developed. As with the other phases of the overall process, we need to think more systematically about implementation and to integrate it into the other phases.[5] Certainly, one must understand implementation mechanisms before government (and other) performance can be evaluated and improved—the next step in the sequence. Initiation or invention and estimation are primarily forward-looking activities. Selection emphasizes the present. *Evaluation* is basically backward-looking; it is concerned with inquiries into system performance and individual responsibility, and is confined to determining how well problems are being dealt with and resolved. Typical topics and questions reflected in the concept of evaluation include: What officials and what policies and programs were successful or unsuccessful in resolving a given

---

[4] The discipline of political science has concerned itself with selection; indeed, much of what political scientists do is in one way or another related to clarifying, measuring, and understanding this phase of the process. One durable and comprehensible summary statement can be found in Robert A. Dahl and C. E. Lindblom, *Politics, Economics, and Welfare* (Chicago: University of Chicago Press, 1976 ed.).

[5] Erwin C. Hargrove, *The Missing Link: The Study of the Implementation of Social Policy* (Washington, D.C.: The Urban Institute, 1975), is representative of heightening interest in this phase of the process and illustrates many of the key concerns and questions as well.

problem? How can one assess and measure performance? What criteria were used to make those determinations? Who made the assessment, and what were the assessor's purposes? Evaluation is a necessary input to the next and final phase of the process.

Termination is necessary when policies and programs have become dysfunctional, redundant, outmoded, unnecessary, and so forth. This phase is not a well-developed one conceptually; however, its importance should not be rated by our lack of understanding of it—as has been startlingly illustrated by recent taxpayer initiatives such as California's Proposition 13, and by a growing concern in the federal legislature for "sunset laws" and "zero-based budgets."[6] How, for instance, can a policy or program be adjusted or terminated without having been thoroughly evaluated? Who suffers as a result of termination? What provisions for redress have to be considered? What costs are involved in termination? Can they be met? What can be learned from termination that can be applied to the initiation and invention of new policies and programs in the same or related fields? The list of relevant questions is long, and neither these questions nor the fact that termination is linked intimately to other steps in the policy process should be ignored.[7]

The value of a conception of a decision-making process is manifold, and several of the possible benefits of this concept are enumerated in Table 1.

Table 1. Possible Benefits of the Policy Process

Phase	*Possible Benefits*
Initiation/ Invention	• Recognition of a problem • Creative thinking about a problem • Prototypical design • Crude hypothesis testing • Preliminary investigation of concepts
Estimation	• Scientific examination of likely impacts and outcomes of plausible options • Normative/evaluative examinations • Development of outlines of a complex policy including program details

[6] See Garry D. Brewer, "Termination: Hard Questions, Harder Choices," *Public Administration Review* 38 (July/August 1978): 338-344.

[7] The entire issue of *Policy Sciences* 7 (June 1976) is devoted to termination topics and practices.

- Appraisal of claims of key participants
- Development of performance indicators
- Specification and estimation of key parameters

Selection
- Focusing of debate on actual issues
- Allowance for political compromises
- Realistic choice among program designs and options
- Reduction of uncertainties about options

Implementation
- Development of rules, regulations, and guidelines for execution
- Development of specific pieces of a program
- Establishment of performance standards based on previous estimates
- Minimization of execution costs

Evaluation
- Comparison of estimated and expected performance levels with those attained
- Assignment of responsibility and sanctions

Termination
- Determination of whether problem is chronic, recurring, or resolvable
- Generation of information about new problems created in termination acts, some of which may require planning, political attention, and treatment

Marine Policy and the Process: An Overall Assessment

With respect to the idea of a "life cycle" for the decision or policy process, several summary observations about marine and ocean policy are worth making.

With the exception of the selection phase, there are few, if any, identifiable institutions or individuals primarily concerned with the various phases of the process. Problems have been identified from the earliest days of the Republic, usually by persons who have studied and who have become concerned about the historic aimlessness and the potential consequences of ad hoc policies and programs; however, the warnings have not been numerous nor have they been acted on with even a fraction of the energy and resources that would be needed to "make a difference." In short, initiation and invention have been sporadic and have generally been far removed from the sources of power and wealth which could and should have been galvanized to make creative and useful decisions. The task of estimation has been grossly undercapitalized and neglected. It is remarkable that centers for the study of ocean management problems have only been created within the past decade. Scholarship and analysis are likewise in scarce supply. Implementation has been left to a bewildering array of offices and agencies, and the results have been far from satisfactory—a point made in indictments which come periodically from the U.S. General Accounting Office (GAO), a primary source of evaluation in this area. Apart from the GAO, however, there are few examples of systematic, comprehensive, and authoritative evaluations; this phase of the process is embryonic at best. Termination, with the notable exceptions of involuntary cessations resulting from mismanagement and overfishing, has not even been considered, much less institutionalized in a responsible way.

This is not to say that the federal government should be held accountable for these glaring deficiencies. Nor is it to say that future efforts should stress a concentration of attention and effort at the federal level. Rather, a full flowering of ocean management initiatives in a variety of governmental and other institutional settings is long overdue. Lacking such developments, the current system suffers from an overemphasis on political choices made without benefit of consideration of ocean problems from a variety of perspectives. In fact, special interests appear to have dominated the process so far. It is a system that fails to learn, from its own mistakes or from mistakes made in closely related fields, e.g., environment, national security, and others. And it is a system that is headed for monumental difficulties if many concerned scholars, analysts, and officials do not begin to face these facts.

Any one of the individual phases of the decision-making process could be singled out for more detailed attention, especially such areas as the information, resources, demands, and institutions that are operating or that could be devised to improve current prevailing practice. However, the remainder of this discussion concentrates on those phases of the process that appear to be the most critical with respect to marine policy.

While the details of the following assessment may be disputed, its basic message is hard to discount.

> Despite all the efforts of the American people, the Congress and the President to focus on a national ocean program, despite the several acts and administrative reorganizations within the federal government, and despite the extraordinary resources channeled to scientific research, surveys, vessel construction, subsidies, and other

forms of aid to the marine community. . . .it was doubtful whether the resources of the eleven departments and agencies closely involved with marine affairs "are being applied to best serve national purpose." What troubled many observers of the development of American policy for the oceans was the fragmentation of the decision-making process into several agencies, often competing or overlapping in their functions. Washington, beset not only by the organized interests of the shipping industry, fishermen, energy producers, environmentalists, and others, but also traumatized by the politics of its bureaucracies seemed inept in setting priorities and incapable of implementing a strong, purposeful ocean policy to embrace both domestic and international needs.[8]

From this assessment, several severe weaknesses of the existing decision-making process tend to stand out, and require comment and attention.

Severe Weaknesses in the Process

Initiation

Much decision-making activity in marine affairs during the last decade appears to have been conducted without benefit of suitable or adequate information and intelligence about the nature of the problems being confronted. For example, the Fisheries Conservation and Management Act (P.L. 94-265) mandated revolutionary changes in policies and programs, but did so mainly in response to strident appeals from those with special interests and without much regard to basic facts about the context or setting in which these changes were to take place.[9] Notwithstanding the desire to protect American fishermen, one cannot determine from the Act the meaning of such terms as "maximum sustainable yield," "optimal sustainable yield," "harvesting capacity," or a host of other fundamental terms and provisions. Abdicating responsibility for these determinations to regional councils not only guarantees uneven and contradictory resolution of the matter, but it also stresses the thoughtlessness of those responsible for the formulation of

---

[8] Gerard J. Mangone, *Marine Policy for America* (Lexington, Mass.: D.C. Heath, 1977), pp. 39-40. (Internal reference is to U.S. GAO, *The Need for a National Ocean Program and Plan,* which is a rare effort and example of evaluation.)

[9] Again, an evaluation-based study seems to have captured many of the key and still unresolved problems and dilemmas: U.S. General Accounting Office, *The U. S. Fishing Industry: Present Condition and Future Condition of Marine Fisheries* (Washington, D.C.: Government Printing Office, December 1976).

the Act (and its attendant policies and programs). It is, in effect, decision-making without specification; policy-making without information.

Contextual specification requires concentration on several key elements: (1) the identification of criteria to define the problem; (2) the specification of relevant environmental parameters; and (3) the identification of the time frame and other constraints that help to bound the problem.

The identification of proper criteria is fundamental to policy and planning tasks. Unless the appropriate questions are asked, it is virtually impossible to obtain, or even recognize, meaningful answers. Ed Quade put the matter nicely when he wrote, "An analysis must begin with problem formulation. A major pitfall is the failure to allocate the total time intelligently, so that a sufficient share of it will be spent in deciding what the problem really is."[10] A common error in identifying criteria is to define the problem based either on inadequate information or solely upon existing conditions in the problem setting. Another danger would be to base criteria on symptoms rather than on underlying causes. All of these difficulties appear to exist with respect to the Fisheries Conservation and Management Act.

The fisheries probem, stated most simply, is not one of providing protection but rather one of changing the fundamental composition and characteristics of fisheries practice and management. If the General Accounting Office is correct in its assessment that American coastal fishing is under-capitalized, inefficient, uneconomical, and resistant to technological innovation,[11] among other things, then this Act not only does not answer these problems, it contributes to them and encourages their proliferation. The problem and the solution that was chosen are both misspecified.

A responsible analysis of the problem would have resulted, after identification of key criteria, in delineation of the relevant actors, a determination of what their interactions are and can be expected to be, and an assessment of how the problem can be bounded and analyzed.

The identification and assessment of actors and clients, both those who stand to benefit and those who inevitably lose, is an overtly political activity that continues throughout the entire decision process. However, the analyst is responsible for defining who is likely to be affected by a particular problem and the alternatives that might be generated in its solution; without such information, no one can judge the importance or priority of the problem. One common error at this step has been to define

---

[10] Edward S. Quade, "Pitfalls and Limitations," in E.S. Quade and W. I. Boucher, eds., *Systems Analysis and Policy Planning* (New York: American Elsevier, 1968), p. 75.

[11] U. S. General Accounting Office, *The U.S. Fishing Industry.*

marine problems in local, regional, or national terms. It must be emphasized that most marine issues are, in fact, worldwide in scope and have numerous interdependencies that can be ignored only at great peril.

Bounding the problem, the final part of problem definition, requires that the analyst reduce the problem to manageable proportions; the analyst has only limited time and information, and choices must be made. In a very few cases, the whole problem may be dealt with directly; more often, the problem must be broken down into a number of smaller and more manageable sub-problems. What appears to have happened in the fisheries conservation case is that only one small aspect of the problem was treated directly within the policy and programs created in the subject act. The concern for protection has dominated the time, attention, and deliberations of those responsible to the point where basic scientific matters, such as determining various types of yield, the size and disposition of fish stocks, and the current and desirable means of investing in and operating in the coastal fisheries have all been slighted.

Estimation

Fisheries research and analysis provide many specific examples of shortcomings in the analytic components of the decision process that is currently utilized for shaping marine policies.[12]

For those few areas in the world where basic statistics have been collected, such as the North Sea fishery, the identification, assembly, and maintenance of data have proven to be formidable, time-consuming, and expensive tasks.[13]

> It is a matter of history that, for the best-known and most studied fishery in the world, it has taken approximately fifty years to introduce its first international regulation, only to find that there are still many problems to solve. What is important, however, is that it is now possible to foresee some of the dangers as well as to predict with some certainty the prospective benefits. This is only because there are now available long series of statistics, steadily improving in precision for conservation purposes, *albeit still not entirely adequate.*

[12] Several of these ideas were developed with the considerable assistance of colleagues laboring together in a working party to elaborate *The Scientific Bases for Determining Management Measures.* Fruits of these labors, available under this title, can be obtained from the United Nations Food and Agriculture Office in Rome, Italy: FAO Fisheries Report No. 236, FIRM/R236 [En], 1980.
[13] Cyril E. Lucas, "Regulation of North Sea Fisheries under the Convention of 1946," *Rome Papers* (1955), pp. 167-177, as cited in McDougal and Burke, *The Public Order of Oceans,* pp. 462-463. (Emphasis added)

For the most studied fishery in the world, more than seventy years and unknown thousands of man-hours have been expended in the collection of basic information, the utility of which for policy and management purposes is still subject to dispute.

Progress in primary data collection has been remarkably slow, and there is little reason to believe that this will change in the foreseeable future. Inadequate research resources are being expended for these purposes,[14] and the complexity of the subject matter has also limited progress in this area.[15] Indeed, fishery management appears to be confronted with a near classic, circular problem: We have little usable theory because there is so little solid information and data; but there are few decent data because there have been so few good analyses to generate theoretical insights indicating the data to collect. It is a problem made all the more difficult by the related facts that investment capital, credit, and insurance are all scarce, and hence limit the exploration and development of new fisheries. The result is that solid, empirical information about the true potential of the world's fish resources is slow to accumulate.

One general result of the data problem is the increasing demand being placed on the few scientists in the field to provide definitive, if not authoritative, answers to contentious policy questions. When, as must necessarily be the case, a policy-maker's query is met with equivocal and guarded replies derived from tentative and uncertain models of systems only vaguely understood, the prevalent tendency to dismiss or overlook scientifically based advice becomes all the more understandable. Furthermore, such a response even begins to look rational when the scientists argue among themselves about even the most fundamental concepts and measurements. For instance, the concept of maximum sustainable yield (MSY) has received much research attention, primarily because it spotlights conservation goals shared by many fishery biologists; however, even within this group solid consensus still does not appear to have been attained, and the issue is far from resolved.[16] Outside of this

[14] One is impressed, on reading any representative collection of fishery studies, by the small scale and narrowly biologic character of the questions and researches that are being asked and undertaken, as compared with the large-scale and more-than-biologic character of the management and policy problems that fisheries represent. R. R. H. Harden-Jones, ed., *Sea Fisheries Research* (London: Elek Science, 1974), is exemplary in this regard.

[15] C. W. Clark, *Mathematical Bioeconomics* (New York: Wiley Interscience, 1976) is quite a nice departure from the usual in terms of its explicit consideration of several of the more relevant complexities of this system.

[16] A reading of International Council for the Exploration of the Sea, *Cooperative Research Report* (Report of the Ad Hoc Meeting on the Provision of Advice on the Biological Basis for Fisheries Management, Charlottenlund, Denmark, February 1977), can only lead to this interpretation of the current state of affairs.

group, particularly with respect to economists and politicians, the concept and the goal that MSY represents are even less generally understood and embraced.[17] Indeed, since scientific analyses do not deal with such issues as fishing industry economics (including profitability and technological innovation), the quality of a fisherman's life, the optimal use of fish resources, and the provision to the consumer of a steady, dependable supply of fish at a reasonable cost, one must conclude that such analyses have misspecified the actual problems confronting decision-makers in the field.

Selection

Not surprisingly, decision-making in fisheries has historically evolved piecemeal and without compelling or thoughtful analysis. Decisions, where they have been made at all, have been reached in a crisis atmosphere and have been taken to protect special local interests during the final phases of stock depletion. They have been short-sighted and patently unmindful of longer range consequences;[18] they have substituted bogus conservation arguments for actual economic ones;[19] and they have shown precious little concern for the related objectives of open access to a primary resource, identification and maintenance of stock, and equitable economic return for fishing effort.[20] According to Mangone, with the notable exception of the International Whaling Commission, "which has had the responsibility for a species on a world-wide basis, the activities of the commissions have been generally limited to 'convention' areas and a few species." Such activities have also been limited to

[17] The economic critique is variously spelled out in Lee G. Anderson, ed., *Extended Fisheries Jursidiction* (Ann Arbor: Science Press, 1977), where arguments for well-balanced theory combining economic, social, and political elements with the more restricted biological ones are presented; the practical and political critique is presented, often brutally, in U.S. General Accounting Office, *The U.S. Fishing Industry.*

[18] Francis T. Christy and Andrew Scott, *The Common Wealth in Ocean Fisheries* (Baltimore: Johns Hopkins University Press, 1965). Chapter 2 provides a capsule summary indicative of this contention.

[19] See D. H. N. Johnson, "Developments since the Geneva Conferences of 1958 and 1960; Anglo Scandinavian Agreements Concerning the Territorial Sea and Fishing Limits," *International and Comparative Law Quarterly* 10(1961):587, 590-91. The reference is to Iceland.

[20] Douglas M. Johnston, *The International Law of Fisheries* (New Haven: Yale University Press, 1965); and Albert W. Koers, *International Regulation of Marine Fisheries* (London: Fishing News Books, 1973), both make these points in considerable detail, especially with respect to decisions made about tuna, whales, and halibut.

data collection and research, as a base for recommending measures for conservation and development. But only the halibut, salmon, and tropical tuna commissions have had their own international staff, with fairly modest budgets. And in these three instances the budgets were almost entirely supported by the United States and Canada.

And, the recommendations of these commissions, often presented without definitive sanctioning power, have concentrated on conservation measures associated with "closed seasons, gear requirements, and catch quotas."[21] Conservation recommendations, furthermore, have been dominated by biological, not economic or institutional, concerns and values.[22] At least one critical observer of the Pacific halibut industry has gone so far as to point out that a "successful" conservation effort has, in fact, been economically and socially wasteful,[23] a contention made more generally and forcefully by H. Scott Gordon, who stated:

> It is my belief that most conservation policies that are now in operation fail to achieve a rational economic objective and many of them are detrimental to the progress of the fishing industry, the incomes of fishermen, and the welfare of society.[24]

It would be surprising if comparably discomforting arguments could not be made with respect to other key elements of the marine policy environment. It is also quite understandable to see, in the absence of any guiding conception of decision-making as an integral process, that the implementation of marine initiatives has also been less than satisfactory.

Implementation

Implementation does not "just happen" as an aftermath of political or other decisions. It is generally a lengthy, interactive process in which the initial "solutions" embedded within the decision are tried out, tested, and changed in response to the realities of the situation. Perhaps far too little is known about how marine policies and programs have been implemented in the past. As a result, insufficient attention has been paid by decision-makers to the feasibility of the options they consider and

[21] See Mangone, *Marine Policy for America,* p. 252, for all of the quoted summary assessments.
[22] Francis T. Christy, Jr., "New Dimensions for Transnational Marine Resources," *American Economic Review, Papers and Proceedings of the 82nd Annual Meeting,* 2,60 (May 1970): 109-113.
[23] James A. Crutchfield and Arnold Zellner, *Economic Aspects of the Pacific Halibut Fishery* (Washington, D.C.: U. S. Dept. of Interior, Fishery Industry Research Series, Vol. 1, 1961).
[24] As cited in McDougal and Burke, *The Public Order of the Oceans,* p. 482.

select. A choice that may be "politically optimal," in the sense that a consensus can be built and sustained, may not necessarily be optimal, much less feasible, from the point of view of those charged with its execution. An infeasible option is no option at all.

Several questions help to make this point: How have past decisions related to the marine setting been implemented? What have been typical responses by those whose lives and livelihoods are changed as a result of these decisions? What kinds of data and information have been collected and are needed to make these determinations? Are they worth collecting; have they been collected; have they improved subsequent decisions and modifications to existing and prevalent ones? Not much is known about any of this, and it would seem that most of these questions have been unasked and unanswered too long.

Implementation as a phase in the decision process needs to become better understood if we are to begin the necessary intellectual and practical tasks of improving marine policies and other public policies as well.

For instance, who is responsible for the decision—what is the source of the policy? It might be a presidential directive (various reorganization plans affecting marine affairs), the passage of legislation (Outer Continental Shelf Lands Act of 1953), simple administrative decisions (allocation of the U. S. Coast Guard's annual budget to various activities), or actions of one of several types of courts or international bodies (Law of the Sea Conferences). The point is that each source has different roles and functions that often determine how a policy is defined, selected, and implemented. In the case of America's Merchant Marine Act of 1970 (P. L. 91-469) the fact that federal legislation is the most obvious source of the policy is essential, but it is not the only fact that must be considered. For instance, was the law passed in the wake of extensive debate and consideration of the various provisions—careful reading of the progress of the legislation indicates that this did not occur[25] —or did a relative handful of representatives of powerful special interests tailor the decision to their own needs? Did it have narrow or overwhelming support? If support was narrow, did this mean that the wording of the legislation was left intentionally vague in the interests of building consensus and passing the law? How are the actions of other policy sources, e.g., local courts, national boards, or international commissions, likely to reshape the policy through time? In the case cited, it appears as though effective implementation was rendered nearly

---

[25] U. S. House of Representatives, Committee on Merchant Marine and Fisheries, *Hearings,* "President's Maritime Program," (Washington, D.C.: Government Printing Office, No. 91-17, 1970).

impossible, and a careful assessment of the source, including the incentives and motivations of the key participants, could have led one to this conclusion without undue difficulty very early.[26]

The complexity of the administrative process is another general implementation concern that demands consideration. This refers to the number and interrelationships of different agencies that must be coordinated—both horizontally and vertically—in order for a decision to be implemented. Obviously, the greater the number of institutions that have to be considered, the more difficult and complex the process of implementation.[27] And generally, the further removed the decision-maker is from the implementing agent and the client, the greater the opportunity for distortions of the policy or for variations from original intent. The Commercial Fisheries and Research Development Act of 1964, as amended, illustrates some of what is at issue here. In the words of Mangone,

> The overblown expectations of invigorating the American fishery industry while assisting the developing nations of the world were thwarted by the initial high costs of manufacture, wrangling within the industry and government, bureaucratic inertia, and weak responses in the undernourished countries. Least helpful was the conservative attitude of the U. S. Food and Drug Administration, which first regarded the crushed fishmeat and bones as "filthy" and then as possibly dangerous with their residues of lead and fluoride.[28]

This calls attention to the essential question of just how involved the high-level decision-maker should become to ensure that the project is implemented as intended. One would not expect such individuals to become involved in the day-to-day details of implementations; a more parsimonious strategy for these individuals is to stress evaluation, careful specification of intent, and thorough consideration of the motivation and incentives of those called on to implement the details of the program. In this case, such an appreciation and style were absent, leading to the conclusion that "whether the total investment of public revenues into fisheries had justified the return to the economy of the nation was debatable."[29]

[26] Charles L. Schultze, "The Role of Incentives, Penalties and Rewards in Attaining Effective Policy," in Julius Margolis and Robert Haveman, eds., *Public Expenditures and Policy Analysis* (Chicago: Markham, 1970). The idea of implementation as a "game" is stressed well in Eugene Bardach, *The Implementation Game* (Cambridge, Mass.: M.I.T. Press, 1977).

[27] Todd La Porte, ed., *Organized Social Complexity: Challenge to Politics and Policy* (Princeton: Princeton University Press, 1975), is replete with cases and ideas on these and closely related issues and problems.

[28] Mangone, *Marine Policy for America*, p. 137.

[29] *Ibid.*

Although this short discussion deals with only a few of the factors influencing implementation, hopefully it points the way to some of the missing ideas and analyses in the marine policy arena. Nevertheless, the discussion is far from comprehensive. Implementation is non-trivial, and an appreciation of the enormity of the proper implementation task would help in understanding why analysts and practitioners have addressed the topic less than enthusiastically. Decision-makers are equally reluctant to address the substantive issues of implementation, and the act of merely issuing a new set of rules and regulations in the face of past failures only neglects central problems of implementation: rules are subject to a variety of interpretations by individuals throughout the system, and plans generally fail to account for the unanticipated as well as for organizational and personal obstacles that can and almost certainly will arise.

The Evolving Task

It is somewhat reassuring to realize that many of the problems inherent in a full appreciation and understanding of the decision processes responsible for marine policy are beginning to surface and to demand attention and clarification. The creation of centers devoted to the study of marine policy within the United States is one helpful sign; likewise, one must take some small measure of comfort in the initiatives of many international bodies to cope with international aspects of the marine environment, such as the Exclusive Economic Zones (EEZ) created within the terms of the Law of the Sea Conference. However, one must stress not only the incredible tasks of institution building, conceptualization, and analysis that face those for whom the world's oceans and waterways have interest and appeal, but also the enormity of the stakes at issue. As compared with the resources expended for technical analyses of scientific marine questions, the sum so far expended on marine policy formulation and analysis is paltry indeed. (This judgment is made on an impressionistic basis, given that no proper accounting of both general classes of expenditure has been made.) As compared with the stakes, the little time, diffuse attention, and scant resources being devoted to marine policy are negligible. The time is long overdue when all best efforts should have been bent toward the formulation and implementation of comprehensive national and world ocean programs—efforts that may only begin after the full complexity of the decision-making process is appreciated and, in time, mastered.

CHAPTER 7
MUDDLING-THROUGH, MODELING-THROUGH, COMPREHENSIVE-INTEGRATING, AND UNITARY-RATIONAL-ACTOR POLICY-MAKING PROCESSES

FRANCIS W. HOOLE*

Introduction

As marine and coastal zone matters grow in importance and salience so do the complexities of issues, programs, policies, and agencies formed to handle them. The ocean policies and programs of governments deal with an enormously complex world where information is tentative and incomplete and yet is frequently so abundant as to overwhelm policy-makers. These policies and programs are the result of highly political policy-making processes involving numerous individuals with varying and often conflicting interests.

In this chapter, I will focus on the governmental ocean policy-making processes and attempt to sort out the management science techniques and systems that reduce uncertainty and assist policy-makers in dealing with complexity.

Ocean Policy-Making Processes

A policy-making process consists of the series of events involved in making and executing decisions.[1] All policy-making processes dealing with

* Sections of this essay originally appeared in a paper entitled "From Muddling Through to Modeling Through," which was presented at a conference on Formulating Marine Policy: Limitations to Rational Decision Making, sponsored by the Center for Ocean Management Studies, University of Rhode Island, in cooperation with the Office of Ocean Management and Office of Sea Grant of the National Oceanic and Atmospheric Administration, Narragansett Bay Campus, University of Rhode Island, June 19-21, 1978. Support for this chapter was provided by the Institute for Marine and Coastal Studies at the University of Southern California. Naturally the final product and any errors contained in it are my responsibility. (Francis W. Hoole is Professor of Political Science at Indiana University.)

[1] This general orientation is developed more fully and used in an analysis of budgetary activities in an international organization in Francis W. Hoole, *Politics and Budgeting in the World Health Organization* (Bloomington, Ind.: Indiana University Press, 1976). Building upon that study, L. Harmon Zeigler and Harvey J. Tucker have conducted a somewhat similar analysis of American state and local politics. See L. Harmon Zeigler and Harvey J.

marine and coastal matters have certain things in common. Each is essentially a political process where policy-makers engage in a struggle over proposed actions. Most ocean policy-making processes are highly structured and complex, and are viewed usefully as cybernetic systems. Inputs consist of information that comes into the policy-making process and is transformed by it. Outputs are the products of the system and they are called actions. The system is the mechanism that contains the policy-making process. It transforms inputs into outputs. Feedback consists of information regarding the results of actions.

A decision is a choice among alternative courses of action. An action is a deed or behavior. It is useful to conceive of a hierarchy of actions. Lower-level actions involve day-to-day activities, and there are clearly defined and agreed upon single objectives. These are *operational actions*. A middle-level action involves multiple and often conflicting objectives but an established overall policy which narrows the focus to the choice of alternatives for implementing the policy. These are *strategic actions*. The highest level of action involves decisions with conflicts in both objectives and guiding policies. These are *policy actions*.[2]

The participants in the policy-making process evaluate inputs and decide upon governmental actions. It is probably sufficient to mention seven major categories of ocean policy-makers: (1) legislators, who are usually elected representatives and serve as members of the parliamentary bodies of the government (e.g., senators, city councilmen); (2) the executive head, who is usually elected and is responsible for the administrative operations (e.g., presidents, governors, and mayors); (3) bureaucrats, who are civil servants and comprise the governmental bureaucracy; (4) judges, who provide legal interpretations; (5) representatives of other governments, who are probably most frequently bureaucrats for governments involved in joint marine and coastal activities; (6) representatives of nongovernmental organizations, who most frequently represent interest groups and businesses; and (7) experts, who are selected because of their technical expertise and usually serve in their personal capacity.

These policy-makers consider a variety of information when deciding upon and implementing marine and coastal activities for governments. In interacting in the policy-making processes, which are, of course, highly

Tucker, *The Quest for Responsive Government* (North Scituate, Mass.: Duxbury Press, 1978), pp. 135-87. This same orientation also serves as the basis for the analysis of data from three international organizations in Francis W. Hoole, Brian L. Job and Harvey J. Tucker, "Incremental Budgeting and International Organizations," *American Journal of Political Science* 20 (1976):273-301.

[2] These types of action are identified and described more fully in Jack Byrd, Jr., *Operations Research Models for Public Administration* (Lexington, Mass.: Lexington Books, 1975), pp. 11-13.

political, the ocean policy-makers develop patterns of behavior which I will call policy-making rules. A policy-making rule is the calculus by which information coming into a policy-making process is transformed into policies and programs.

The events involved in deciding and implementing ocean policies and programs usually take place over a period of several months. It seems adequate to assume that governmental ocean policy-making processes involve cycles with the following six steps or subsystems: (1) the proposal development subsystem, which involves preparation of specific proposals for action (such as budget proposals by a department head); (2) the executive head subsystem, which involves strategic and policy decisions by the executive head; (3) the legislative subsystem, which involves policy-making by the parliamentary bodies of government; (4) the supplementary change subsystem, which involves changes in decisions; (5) the implementation subsystem, which involves carrying out the decisions; and (6) the evaluation subsystem, which involves the evaluation of activities carried out by a government and the feedback of this information for subsequent policy-making. Each of the subsystems occurs during a different period in time, although there is overlap, and each involves distinct governmental activities. Each subsystem involves a slightly altered set of circumstances with differences existing in inputs, activities of policy-makers, and policy-making rules.

I would like to discuss four versions of governmental ocean policy-making processes. These conceptualizations fall along a continuum in the following manner:

Muddling-	Modeling-	Comprehensive-	Unitary-
Through	Through	Integrating	Rational Actor
Policy-making	Policy-making	Policy-making	Policy-making
Processes	Processes	Processes	Processes

As movement takes place from left to right, there is an increase in the reliance on systematic management science methodologies.

Muddling-Through Policy-Making Processes

One of the most striking features of the muddling-through (or bureaucratic politics) conceptualization is its viewpoint regarding complexity.[3] Numerous policy-makers representing varieties of viewpoints

---

[3] For elaboration of this conceptualization, see Charles E. Lindblom, "The Science of Muddling Through," *Public Administration Review* 19 (1959):79-88; Graham T. Allison and Morton H. Halperin, "Bureaucratic Politics: A Paradigm and Some Policy Implications,"

are dealing simultaneously with various overlapping issues. The policy-making process is overwhelmed by cross-cutting political currents and complicated task environments. This highly political process is characterized by bargaining involving complicated tradeoffs and is highly decentralized with coordination taking place through elaborate forms of partisan mutual adjustment. Information is incomplete but so abundant as to be overwhelming. Policy-makers reduce uncertainty and ambiguity and adapt to changing environments by decomposing complex problems into more easily handled subproblems. They use simplifying rules of thumb, and the policy-making rules tend to be dynamic, serial in nature, satisficing, disjointed, incremental, bounded in terms of rationality, definitely not ends-means oriented, and frequently differ according to subsystem, issue-area, and level of action. There is little role for management science techniques.

Policy-making processes of the muddling-through type appear to be fairly typical in the marine and coastal context. But while many observers feel that these characterizations present accurate descriptions of how policies are made, many individuals question whether they provide a description of the way policies should be made.

Modeling-Through Policy-Making Processes

It is the role of management science modeling techniques that distinguishes the otherwise similar muddling-through and modeling-through policy-making processes.[4] Various management science analysis techniques are used in a modeling-through process, and the results of these problem-specific studies are used in a partisan manner. Thus an attempt is made in a modeling-through process to systematize somewhat the selection and processing of problem-specific information and richer, more focused, and more rigorous analyses are undertaken in an effort to supplement those normally found in a muddling-through policy-making process.

World Politics 24 (1972), Supplement 40-79; and John D. Steinbruner, *The Cybernetic Theory of Decision* (Princeton, N.J.: Princeton University Press, 1974), pp. 47-139.

[4] I believe Bertram M. Gross was the first to use the phrase modeling-through. See Bertram M. Gross, "Management Strategy for Economic and Social Development: Part II," *Policy Sciences* 3 (1972):14. Unfortunately, he did not develop very fully the definition or meaning of a modeling-through policy-making process. The meaning attributed in this essay is mine. For elaboration, see Francis W. Hoole, "From Muddling Through to Modeling Through," in *Formulating Marine Policy: Limitations to Rational Decision-Making, Proceedings of the Second Annual Conference Held at the Center for Ocean Management Studies,* Timothy M. Hennessey, chairman (Kingston, R.I.: Center for Ocean Management Studies, University of Rhode Island, 1979), pp. 80-93.

Several management science techniques are potentially useful for generating information frequently needed by policy-makers working on marine and coastal activities.[5] Cost-benefit, cost-effectiveness, and cost-utility techniques emphasize the calculation of the ratio of cost to return for a policy or program. They are part of a general approach to efficiency.[6] Operations research techniques focus on the derivation of optimal solutions to complex operational problems, although they also may have some usefulness for strategic problems in a wide range of issue areas.[7]

[5] I make no claim to be discussing all of the potentially relevant techniques but I do believe that the techniques that are discussed are among the ones that ocean policy-makers will find most helpful.

[6] Cost-benefit, cost-effectiveness, and cost-utility techniques can be helpful to policy-makers in the marine and coastal context. For example, analyses of the costs and benefits of enforcing the 200-mile fishing regulations would be useful, as would studies concerning the siting of liquid natural gas and nuclear power facilities. For a general introduction to these techniques see: Jerome Rothenberg, "Cost-Benefit Analysis: A Methodological Exposition," in *Handbook of Evaluation Research*, Vol. 2, ed. Marcia Guttentag and Elmer L. Struening (Beverly Hills, Calif.: Sage Publications, 1975), pp. 55-88; Henry W. Levin, "Cost-Effectiveness Analysis in Evaluation Research," in Guttentag and Struening, eds., *Handbook of Evaluation Research,* pp. 89-122; Christopher K. McKenna, *Quantitative Methods for Public Decision Making* (New York: McGraw-Hill, 1980), pp. 127-161; Edith Stokey and Richard Zeckhauser, *A Primer for Policy Analysis* (New York: W.W. Norton & Co., 1978), pp. 134-58; and A. R. Prest and R. Turvey, "Cost-Benefit Analysis: A Survey," *Economic Journal* 75 (December 1975):683-775. For background information concerning cost-benefit analysis and coastal activities, see J. W. Devanney III, G. Ashe and B. Parkhurst, *Parable Beach: A Primer in Coastal Zone Economics* (Cambridge, Mass.: The MIT Press, 1976). It seems to me that these techniques would be most helpful for bureaucrats participating in marine and coastal policy-making processes in the proposal development and evaluation subsystems, although they also might be used by bureaucrats in the supplementary change subsystem to evaluate possible program revisions. Specific cost-benefit, cost-effectiveness, and cost-utility information could also be given to other types of policy-makers in the executive head and legislative subsystems. These techniques are potentially useful across a wide range of issue areas and they probably will be most helpful for strategic level decisions, although they also may have utility for operational decisions.

[7] The operations research techniques include the following: project evaluation and review technique (PERT), critical path method (CPM), simulation, linear and nonlinear programming, network analysis, dynamic programming, inventory models, scheduling procedures, queueing theory, gaming, and replacement analysis. Specific information from studies using operations research techniques could be given to policy-makers in the executive head and legislative subsystems, but it appears that the bureaucrats participating in the proposal development, supplementary change and especially the implementation subsystems would find it most useful. Optimization techniques could be used for numerous allocation problems in the marine and coastal programs. For example, they could be employed to allocate personnel to various tasks with the goal of maximizing port safety or to allocate money to various programs for cleaning up oil spills. PERT/CPM techniques would also be helpful in the management of the implementation of large-scale, complex, time-consuming marine and coastal problems; for example, they might be used to schedule and

Evaluation research techniques are concerned with effectiveness and hence with the full range of operational procedures involved in the systematic empirical examination of hypotheses regarding the actual impact of social action policies and programs.[8] Forecasting techniques provide information about the future by means of extending past trends.[9] At the heart of decision-theoretic techniques is an emphasis on

monitor the construction of a new multiple use park on a remote coast or island. For an introduction to operations research techniques, see Byrd, *Operations Research Models for Public Administration.*

[8] The promise of evaluation research is that it can provide systematic information on the actual impact of activities and feed this information back through the policy-making process in such a way that it can influence the making of subsequent decisions regarding ongoing enterprises. Evaluation research information would be generated in the evaluation subsystem and should be helpful to bureaucrats making operational decisions in the proposal development, supplementary change, and implementation subsystems and to executive heads and legislators making strategic decisions in the executive head, legislative and supplementary change subsystems. The evaluation research techniques are relevant for the examination of a wide range of marine and coastal matters. For example, they could be used to evaluate the impact of the California Coastal Act of 1976 or the extension of certain federal jurisdictions to include 200 miles of coastal waters. For a general introduction to evaluation research, see Peter H. Rossi and Sonia R. Wright, "Evaluation Research: An Assessment of Theory, Practice, and Politics," *Evaluation Quarterly* 1 (1977):5-52; Howard E. Freeman, "The Present Status of Evaluation Research," in *Evaluation Studies Review Annual,* Vol. 2, ed. Marcia Guttentag and Shalom Saar (Beverly Hills: Sage Publications, 1977), pp. 17-51; and Francis W. Hoole, *Evaluation Research and Development Activities* (Beverly Hills, Calif.: Sage Publications, 1978). For information concerning evaluation research and marine activities, see Francis W. Hoole and Robert L. Friedheim, "Evaluation Research and Marine and Coastal Policies," *Marine Technology Society Journal* 12 (August-September 1978):13-17; Francis W. Hoole, "Evaluating Management of Coastal Resources," in *Managing Ocean Resources: A Primer,* ed. Robert L. Friedheim (Boulder, Colo.: Westview Press, 1979), pp. 177-85; Robert E. Bowen, Francis W. Hoole, and Susan H. Anderson, "Evaluating the Impact of Coastal Zone Activities: An Illustration of the Evaluation Research Approach," *Coastal Zone Management Journal* 7 (1980):24-46; and Francis W. Hoole and Robert L. Friedheim, "A Selected Bibliography of Evaluation Research for Marine and Coastal Specialists," Occasional Paper No. 5, Institute for Marine and Coastal Studies, University of Southern California, Los Angeles, California, April 1978.

[9] The forecasting techniques involve the extrapolation of trends using expotential smoothing and autoregressive moving average models, the expansion of detailed budget estimates over future years, examination of hypothetical future scenarios through use of econometric and computer simulation models, as well as other techniques. The forecasting activities are usually carried out by bureaucrats in the proposal development subsystem. The information provided regarding the future probably is most useful for operational and strategic decisions. Forecasting techniques could be used for various purposes by marine and coastal policy-makers. For example, projections of beach attendance could be of assistance in determining the number of lifeguards needed, projections of petroleum availability and use could help in the establishment of policies for offshore drilling, and projections of revenue could be useful in developing budgets. For a general introduction to forecasting techniques, see Steven C. Wheelwright and Spyros Makridakis, *Forecasting Methods for Management,* 2nd ed. (New York: Wiley, 1977).

systematizing and consolidating information on subjective probability estimates of achieving certain end states such as goal attainment. The focus is on the calculation of subjective prior probabilities of items such as program success, which can be recalculated as posterior probabilities after information on program effectiveness is available.[10]

It is worth noting that cost-benefit, cost-effectiveness, and cost-utility, as well as operations research techniques, focus primarily on some aspect of the expenditure of resources. They are of greatest utility on budgetary matters. The forecasting, decision theoretic, and evaluation research techniques are of potential utility in both budgetary and nonbudgetary matters. Each of the techniques is useful in a wide variety of issue areas.

These techniques are complementary, with each being oriented to a different but not inconsistent set of questions. In general, they are most useful to bureaucrats, the so-called middle managers of public sector ocean activities. Success in actually utilizing these techniques appears to decrease rapidly as one moves from operational to strategic to policy actions.

Comprehensive-Integrating Policy-Making Processes

This is a conceptualization of a policy-making process which involves use of management science analysis techniques *and* formal organization for integration of action channels throughout the policy-making process. In comprehensive-integrating processes, the same management science analysis techniques are utilized to examine specific substantive problems as in a modeling-through process. However, in comprehensive-integrating processes there are structural differences in action channels which are aimed at systematically integrating the information and decision flows

[10] Bayesian decision techniques will appeal to those interested in capturing in a formal way the intuition contained in subjective estimates. For example, they might be useful for Sea Grant program managers attempting to assess the probable success of engaging in certain advisory services. These techniques employ a decision-theoretic approach utilizing bayesian statistics and multi-attribute utility analysis. Information obtained using these techniques would be most useful to bureaucrats in handling strategic and policy decisions in the proposal development subsystem, although it might also be helpful for the same types of decisions in the supplementary change subsystem. These techniques would be of equal relevance for policy and strategic decisions by agency chiefs in the executive head subsystem. For an introduction to this approach, see Stokey and Zeckhauser, *A Primer for Policy Analysis,* pp. 201-54; McKenna, *Quantitative Methods for Public Decision Making,* pp. 61-126; Ward Edwards, Marcia Guttentag, and Kurt Snapper, "A Decision-Theoretic Approach to Evaluation Research," in Struening and Guttentag, *Handbook of Evaluation Research,* vol. 1 (Beverly Hills: Sage Publications, 1975), pp. 139-81; and the chapter in this volume by Peter C. Gardiner.

throughout the policy-making process. Hence, some weak form of centralization (or systematic coordination) is attempted through integration of policy-making activities.

The various versions of this conceptualization are based upon or are consistent with a wholistic systems orientation to the policymaking process.[11] Among the major variations are the planning-programming-budgeting system (PPBS), management by objectives (MBO) system, zero-base budgeting (ZBB) system, management information system (MIS), and multiple advocacy system. The planning-programming-budgeting system involves an emphasis in the budgetary process on the type of planning advocated by economists instead of the control emphasis of administrators. Because planning is future-oriented, comprehensive, and means-ends oriented, PPBS has similar characteristics. It focuses on the purpose of activities and provides a system to handle multiple functions of budgeting.[12] The management information system attempts to provide managers with information needed to make decisions.[13] The management by objectives system involves a process that requires that executive heads and bureaucrats identify and monitor progress toward the achievement of objectives in various issue areas.[14] The multiple advocacy

[11] For an introduction to systems concepts, see C. West Churchman, *The Systems Approach* (New York: Dell, 1968).

[12] The PPB system is the intellectual product of a group of economists and systems analysts at the Rand Corporation. There are three essential characteristics associated with PPBS (structural organization, analytical process, and information system) and there are three basic documents which are used in most PPB systems to structure the information for policy-making (program memorandum, program and financial plan, and special analytic studies). The PPB system handles budgetary matters in all issue-areas and ideally is applicable in regard to all subsystems of the policy-making process. In the past PPBS seems to have been most useful for bureaucrats and executive heads in the making of operational and strategic decisions. For an introduction to planning, programming, budgeting systems, see Fremont J. Leyden and Ernest G. Miller, eds., *Planning, Programming, Budgeting: A Systems Approach to Management,* 2nd ed. (Chicago: Markham, 1972); and David Novick, ed., *Program Budgeting: Program Analysis and the Federal Budget* (Cambridge: Harvard University Press, 1965).

[13] The impetus for the MIS movement seems to have been provided by computer scientists and industrial engineers. It is clear that there is no single version of a management information system and there have been numerous MIS failures. Management information systems are relevant for use in all subsystems of the policy-making process and seem to have been most useful for bureaucrats in the making of operational and strategic decisions. For an introduction to management information systems, see Raymond J. Coleman and M. J. Riley, eds., *MIS: Management Dimensions* (San Francisco: Holden-Day, 1973); and McKenna, *Quantitative Methods for Public Decision Making,* pp. 369-92.

[14] The MBO movement got its start in the business world. MBO activity takes place in the evaluation subsystem but through systematic feedback processes would affect subsequent activities in the implementation, supplementary change, executive head, proposal

system is in the tradition of adversary proceedings found in judicial and devil's advocate systems. It calls for a process with no significant maldistribution of resources such as power, influence, competence, information, analytical ability, and bargaining among numerous advocates. It also requires agency chief participation and time for adequate debate.[15] The zero-base budgeting system is a bottom-up one which contains a clear management orientation. It works in such a manner that the lower managers initiate a budget cycle by preparing decision packages and doing an initial ranking. As these proposals travel up through the bureaucracy, they are combined with other decision packages and new rankings are developed at a higher level. This continues until the final rankings by those at the very top and the decision on how many of the decision packages to fund.[16]

development and possibly the legislative subsystems. It provides information that is most useful in the making of operational and strategic decisions. In many ways MBO can be seen as an evaluation research oriented system. For an introduction to the management by objectives approach, see Rodney H. Brady, "MBO Goes to Work in the Public Sector," in *Public Budgeting: Program Planning and Evaluation,* ed. Fremont J. Lyden and Ernest G. Miller, (Chicago: Rand McNally, 1978), 3rd ed., pp. 198-213; Frank P. Sherwood and William J. Page, Jr., "MBO and Public Management," *Public Administration Review* 36 (January-February 1976):5-12; Peter F. Drucker, "What Results Should You Expect? A Users' Guide to MBO," *Public Administration Review* 36 (January-February 1976):12-19; Jerry McCaffery, "MBO and the Federal Budgetary Process," *Public Administration Review* 36 (January-February 1976):33-39; and Harry S. Havens, "MBO and Program Evaluation, or Whatever Happened to PPBS?" *Public Administration Review* 36 (January-February 1976):40-45.

[15] The multiple advocacy system would be located in the executive head subsystem but apparently would control activities in the proposal development, supplementary change, implementation and evaluation subsystems through the use of executive head directives. The multiple advocacy system is useful for the handling of strategic and policy level decisions. Cf., Alexander L. George, "The Case for Multiple Advocacy in Making Foreign Policy," *American Political Science Review* 66 (1972): 751-85; I. M. Destler, "Comment: Multiple Advocacy: Some 'Limits and Costs,'" *American Political Science Review* 66 (1972): 786-90; and Alexander L. George, "Rejoinder to 'Comment' by I. M. Destler," *American Political Science Review* 66 (1972): 791-95.

[16] There are four basics for any ZBB system: (1) decision units; (2) decision packages; (3) the ranking process; and (4) the preparation of the detailed operating budget. The ZBB system encompasses the proposal development, executive head, supplementary change and sometimes the legislative subsystems and is usually linked to the implementation and evaluation subsystems. The executive head and the bureaucrats are the most active policy-makers. The ZBB system is useful for making strategic and operational decisions. For an introduction to the ZBB approach, see Peter A. Pyhrr, *Zero-Base Budgeting: A Practical Management Tool for Evaluating Expenses* (New York: Wiley, 1973); Peter A. Pyhrr, "The Zero-Base Approach to Government Budgeting," in Lyden and Miller, eds., *Public Budgeting: Program Planning and Evaluation,* pp. 235-66; and Andrew B. Fogarty and Augustus B. Turnbull, III, "Legislative Oversight Through a Rotating Zero-Base Budget," in Lyden and Miller, *Public Budgeting,* pp. 371-80.

Each of these systems is relevant across a wide range of issue areas. The PPB, MBO, ZBB, and management information systems are most useful for handling different aspects of operational and strategic actions, while the multiple advocacy system is most useful in making strategic and policy decisions. Each of the systems benefits bureaucrats and executive heads more than other policy-makers. The PPB and ZBB systems are useful in various aspects of the allocation and expenditure of resources. The MBO, multiple advocacy, and management information systems are useful for both budgetary and other matters.

These systems are complementary in many ways. The PPB, ZBB, and multiple advocacy systems are compatible with and can be supplemented by a management information system. The PPB and ZBB systems are compatible, as the recent experience in the federal government has demonstrated. The multiple advocacy system, with its orientation toward top leadership, nonbudgetary, and nonroutine matters would seem to be compatible with the PPB, ZBB, and management information systems with their orientation toward middle management, budgeting, and other routine matters. Merging these systems would create a process that is more similar to the unitary-rational-actor policy-making process than are any of the processes discussed previously.

Unitary-Rational-Actor Policy-Making Processes

This conceptualization of the policy-making process views government as a unitary actor that makes decisions by identifying the relevant goals and objectives, considering the possible alternative courses of action, estimating the consequences of each option, and choosing the alternative which maximizes the prospects of achieving the government's goals and objectives.[17] Graham T. Allison describes such a policy-making process in the following way:

> Happenings . . . are conceived as actions chosen by the . . . government. Governments select the action that will maximize strategic goals and objectives.[18]

> The . . . government, conceived as a rational, unitary decision maker, is the agent. This actor has *one* set of specified goals (the equivalent

[17] Cf., Graham T. Allison, *Essence of Decision: Explaining the Cuban Missile Crisis* (Boston: Little, Brown & Co., 1971), pp. 10-38; and Steinbruner, *The Cybernetic Theory of Decision*, pp. 25-46.

[18] Allison, *Essence of Decision*, p. 32.

of a consistent utility function), *one* set of perceived options, and a *single* estimate of the consequences that follow from each alternative.[19]

Action is chosen in response to the strategic problem the . . . [government] faces.[20]

[The action] . . . is conceived as a steady-state choice among alternative outcomes (rather than, for example, a large number of partial choices in a dynamic stream).[21]

Hence, the policy-makers are viewed as making rational calculations and the policy-making rules involve optimization strategies for the entire government.

How is the rational actor ideal adjusted to a pluralistic democratic process with numerous actors? According to John D. Steinbruner:

The shift from an individual level of analysis (where most of the intellectual development has occurred) to a collective level of analysis is achieved by requiring that the collective process be constrained by an explicit set of calculations, shared by the individuals involved, which meet the analytic criteria advanced at the individual level.[22]

Hence, centralization is required for this type of policy-making process.

The unitary-rational-actor policy-making process ignores complexity resulting from levels of action, a plurality of policy-makers, and decomposition into subsystems and issue areas. Information and its processing are not problems, and the rational policy-making rules are optimizing ones that employ the full range of management science techniques without difficulty. This conceptualization differs significantly from the comprehensive-integrating one with regard to centralization, ease of use of management science techniques, and ease and frequency of rational calculation. Most observers seem to feel that the unitary-rational-actor conceptualization presents an inaccurate description of how policies are actually made, although many individuals feel that policies should be made according to the ideals contained in this conceptualization of the policy-making process.

---

[19] Allison, *Essence of Decision*, pp. 32-33.
[20] Allison, *Essence of Decision*, p. 33.
[21] Allison, *Essence of Decision*, p. 33.
[22] Steinbruner, *The Cybernetic Theory of Decision*, p. 45.

A Perspective on the Types of Policy-Making Processes

What is the likelihood that each of these types of policy-making process will exist in the ocean policy context in the 1980s? It is, of course, impossible to say with certainty. However, it is possible to offer some personal estimates.

There is, in my opinion, a very low probability that United States ocean policy-making processes in the 1980s will resemble the unitary-rational-actor conceptualization. The requisites for this version of the policy-making process, such as centralization, unitary means-ends calculations, ease of use of management science techniques and adequate information, are not likely to be found in the complex and highly political world of ocean policy-making. It would be difficult to locate any unitary-rational-actor ocean policy-making process today, and I suspect that the same will be true in the future.

It is highly likely that versions of muddling-through ocean policy-making processes will be found in the 1980s. This has been the dominant form of policy-making process, and certainly it will continue to exist in the future. However, it may slowly become a process of the past, one that will be seen less and less frequently in the future. It is clear that management science techniques are here to stay, and that they are being used more and more in public policy-making processes, especially at the operational level of action. Hence, in my opinion, it is to be expected that more and more muddling-through policy-making processes will change to similar but more analytically sophisticated modeling-through processes.

It is likely that versions of the comprehensive-integrating ocean policy-making process will be observed in the 1980s. In fact, what we may see in the near future is experimentation, the development, and then partial dismantling or major revision of various versions of this conceptualization of the policy-making process. Some success with this policy-making process can be anticipated at the operational and perhaps at the strategic level of action, but little success seems likely at the policy level. Because this policy-making process utilizes the full range of management science techniques, all of the problems of those techniques will be relevant. In addition, because the development of these systems involves changes in existing action channels and because such development usually systematizes the reallocation of political power, it can be anticipated that resistance to their implementation will be serious. I would estimate that the problems, especially the political ones, are several times greater than those found in modeling-through policy-making processes.

The modeling-through version of ocean policy-making processes is the one likely to be of greatest interest in the near future. It contains the essential characteristics of the highly political and decomposed muddling-through policy-making process, but it recognizes the utility of management science techniques in providing problem specific information. However, these techniques were developed to help policy-makers understand a complex world, and each one employs a complex methodology to achieve this goal. To my way of thinking, a really important question is: Will these techniques really work? To answer this question it is necessary to consider the practical problems in utilizing management science techniques in ocean policy-making processes.

Problems in Utilizing Management Science Techniques

At the risk of oversimplification, I would like to focus upon twelve general types of problems which appear to limit the utility of management science techniques in ocean policy-making processes. I will discuss briefly these problems, without arguing that my list is necessarily a complete one.[23]

Identifying Meaningful Focuses for Analysis. It will not be easy to identify and to state clearly the focuses for techniques that analyze various aspects of marine and coastal activities. There are numerous policies and programs with ambiguous and conflicting goals, unclear strategies concerning implementation, and even general confusion about the basic enterprise. Given the highly political nature of the policy-making processes that produce most marine and coastal policies and programs, it would be surprising to find clarity on many of these issues or even on the purposes for undertaking analyses. Decisions on what to analyze and how to proceed will not be easy ones, and the ground rules may change before the analysis has been completed.

Measurement. The measurement rules used in an analysis provide the crucial link between concepts and data. It is difficult to measure adequately the properties of many concepts, and frequently there is little agreement on how to do so. It is, of course, problematical to conduct public policy analyses involving concepts or their relationships without developing faith in the measurement rules which relate the concepts and data.

---

[23] This section builds on materials presented in Hoole, *Evaluation Research and Development Activities*, pp. 117-70. I am grateful to Robert L. Friedheim for suggesting that the categories entitled communication and personalities be added to my list of potential problems.

Data Collection. It appears there is a great deal of missing data, definitions and categories used for data collection seem to change frequently, many data cover only a short period of time, and oftentimes available data do not lend themselves to providing answers to interesting questions. In brief, there is a shortage of meaningful and reliable data for the analysis of marine and coastal activities. It appears that for the forseeable future, special emphasis will need to be placed on making data collection a part of programs and numerous data collection problems undoubtedly will be encountered as this is done. I have serious reservations about the creation of large-scale expensive data banks without very clearly identified and important uses for the data contained therein. In my opinion, policy concerns or theory should dictate the focuses for analysis and the measurement rules, and hence the data to be collected.

Data Analysis. Problems in analyzing data should be anticipated. Unfortunately, optimal solutions are not yet available for all such problems, although it does appear that reasonable solutions can be found for most of them. Because of the high level of expertise required, it may be necessary for marine and coastal policy-makers to obtain expert advice on data analysis problems. The danger, of course, is that improper data analysis will produce artifacts that result in misleading conclusions.

Ethics of Analysis. There are numerous ethical problems revolving around such issues as: (1) withholding treatment from control group cases; (2) confidentiality of information; and (3) honesty in reporting results. In the final analysis, if the potential ethical problems cannot be resolved then the analysis should not be undertaken. It appears to me that it will be a rare case in the marine and coastal context when ethical concerns will be of sufficient magnitude to force cancellation of an analysis.

Communication. Messages, whether written or oral and whether long or short, will not always be clear because of a variety of factors such as different meanings being given to the same words or phrases. Among the apparent reasons for this would be different orientations and training of individuals. Special care must be taken to avoid jargon and to communicate clearly concerning results of analyses. It should be obvious that even the most persuasive and relevant analyses will have limited impact if the results are not communicated clearly.

Personalities. Personality conflicts or an important individual's personality may place serious constraints on a particular analysis, or, indeed, on the general use of management science techniques. There are no simple solutions for the problems of personalities, although sensitivity and the use of common sense may help a bit in dealing with these problems.

Disentangling the Impact of Specific Activities. There is a problem which will have great salience for some policymakers: How can the impact of a particular agency's activities be evaluated when a program is conducted jointly with other agencies? In these circumstances the activity of the agency must be isolated and an evaluation must be set up. If this is not feasible, then the identification of the separate effect may not be possible.

Organizational Factors. There are numerous bureaucratic and agency related problems that are relevant for practitioners doing policy analyses. I will call these organizational problems. Among the important ones are resistance of bureaucrats to systematic analysis and its implied change, the high level of skill required for using most of the techniques, role and institutional conflicts, and the cost of studies. There also are numerous factors in the task environment of specific agencies which may result in difficulties in using systematic analysis techniques in the ocean policy context. However, while organizational problems will be among the most serious impediments to the use of systems and techniques, it does appear that they can be alleviated somewhat through the use of strategies such as support from important administrators, involvement of bureaucrats in the research enterprise, and clear role definition and authority structure.

Timing of Analyses. If the analyses are not available when they are needed by policy-makers, then they will not be used. It is that simple. The timing of relatively fixed policy-making processes and the extended period of time needed for many analyses mean that the use of systematic methodologies to provide timely information will be constrained. An effort must be made to anticipate well in advance the analyses that need to be conducted, and attempts must be made to adjust these efforts to the constraints imposed by the timing of policy-making processes.

Relevance of Analyses. The analyses must be relevant for the decisions at hand and must be conclusive. If the conclusions are not clear or if the policy ramifications are ambiguous, then the relevance of an analysis is questionable. If the wrong question is addressed, then the analysis will have little effect.

Political Factors. The results of any policy analysis will be only part of the relevant information considered by policy-makers because the findings will be used in political processes where factors other than the results of policy analyses will be of major importance. It is likely that the results will be used frequently in a partisan manner by policy-makers to support positions already arrived at on the basis of partisan considerations. Political factors will be especially relevant to attempts to develop comprehensive-integrating policy-making processes because those attempts will frequently result in shifts in political power. Those in danger of losing power can be expected to oppose changes in the policy-making process.

It can be anticipated that these problems and the complications resulting from their interaction will vary in importance according to analysis and agency. In some instances, one or a combination of these problems may be so severe as to preclude meaningful use of one or several of the techniques. Furthermore, it is to be anticipated that the overall severity of the problems will increase as movement is made from a muddling-through to a modeling-through to a comprehensive integrating policy-making process. However, it is my opinion that many versions of these problems can be solved adequately through a flexible and creative effort on the part of those concerned with the analysis. Indeed, it is difficult to imagine that the problems will be insurmountable in all circumstances. Nevertheless, these problems do present limitations to the use of the techniques, and it seems clear that they are fragile methodologies.

Conclusions and Implications

Assuming that the analysis presented so far is basically correct, then conclusions seem to be in order. The primary advantage of management science analysis techniques and systems is that more systematic and focused information may be made available to policy-makers, especially bureaucrats and executive heads dealing with operational and strategic actions in ocean policy-making processes. However, it would be a mistake to expect too much from these techniques and systems, just as it would be a mistake to dismiss them as having nothing to offer. Management science methodologies appear to be neither as useful as many of their proponents suggest nor as useless as is suggested by many of their opponents.

Because the utility of these methodologies depends on the orientation and training of policy-makers, there is the implication that ocean policy-makers should become thoroughly familiar with them. Although this process has already begun, there appears still to be a need to emphasize courses regarding these methodologies and their use in the marine and coastal context. In fact, training, experience, insight, and time will be required for the meaningful use of management science methodologies in the ocean policy context.

In addition to using these methodologies to study substantive issues, policy-makers and scholars should begin to document systematically the problems of using management science techniques and systems in the making of ocean policy. We still have a great deal to learn about the use of these methodologies, and a contribution could be made by clarification of the problems in moving from muddling-through to modeling-through to comprehensive-integrating ocean policy-making processes.

The emphasis in this chapter has been on sorting out things and presenting a general perspective on policy-making processes and management science methodologies. This perspective should be helpful for both practitioners and scholars concerned with the making of ocean policy.

CHAPTER 8
MULTI-ATTRIBUTE UTILITY MEASUREMENT FOR COASTAL ZONE DECISION-MAKING

PETER C. GARDINER*

Introduction

On August 24, 1979, the MacNeil-Lehrer Report, a program on public television, was devoted to a discussion of a five-year running battle over a proposal to construct a new oil refinery at Portsmouth, Virginia, on the Chesapeake Bay. Involved in the battle and present on the program were high-level representatives of the U.S. Environmental Protection Agency (EPA), the U.S. Department of Energy (DOE), the City of Portsmouth, representatives of the independent oil refinery that wished to build the new refinery, and representatives of the concerned citizens of Portsmouth—a group of Portsmouth citizens formed to fight the proposed refinery construction.

Throughout the discussion on the program, a number of interesting facts were presented as well as many value-laden conclusions. At the time of the broadcast, nine years had elapsed since the initiation of plans for the refinery, and about $20 million had been spent, but no decision had yet been made as to whether or not the refinery could be constructed. The Department of Energy spokesman stated that there was a national energy problem and a critical need for constructing a refinery somewhere on the east coast of the United States. In his view, it was simply a question of which location should be chosen. The elected officials of the City of Portsmouth favored construction of the refinery within its jurisdiction for the subsequent tax advantages to the community. They had conducted a study in which it had been concluded that the benefits outweigh the potential risks. The concerned citizens group opposed the refinery. They cited an EPA study of seventeen potential locations for new oil refineries on the East Coast. That study had concluded that the Portsmouth location was "one of the worst." The EPA, therefore, supported the citizens group while the DOE supported the city officials. Needless to say, the matter was not resolved during the television program. About the only aspects of the decision problem agreed to among the participants were that (1) the process must be speeded up; and (2)

* Peter C. Gardiner is Assistant Professor of Systems Management and Research Associate for the Social Science Research Institute at the University of Southern California.

the ultimate decision should somehow be an amalgamation of the interests of government agencies, the public, the independent refiner, and local governments. None of the participants had the slightest idea how this could be done in practice, and the matter was slowly making its way through the courts.

This situation is typical of many coastal zone decision-making situations. Each involves many impacts and many different and often divergent viewpoints. Coastal zone decisions, such as this one, involve disputes about which course of action is best for the public good. Yet, arguments about which action or policy is best typically hinge on disagreements about value. Such disagreements are usually about degree, not kind. Normally such disagreements—as in this example—drag on and are fought out in adversary proceedings in the context of each specific decision, and so are fought out over and over again, at enormous social cost each time another decision must be made. One can imagine in this example up to seventeen reruns (one for each of the seventeen sites reviewed by the EPA) just for the construction of a single oil refinery on the east coast. Moreover, once a decision has been made there tends to be some suspicion by those involved as to just how the decision was reached and how their own views were incorporated in it, if at all. And this suspicion is a function of what's at stake and the amount of reciprocal ox-goring that took place as well as the perceived loss of face by involved agencies.

Fortunately, technology does exist that can spell out explicitly the values of each individual or group, showing how and how much they differ and frequently reducing the extent of such difference in the process. The technology provides a formal and open mechanism for incorporating these values in a decision-making process. The exploitation of this technology permits regulatory agencies responsible for coastal zone decision making to shift their attention from the specific actions and policies being debated to the values these actions and policies serve and the decision-making mechanisms that implement them. By explicitly negotiating about, agreeing on, and publicizing a set of values, a public agency can, in effect, inform those it regulates about decision-making ground rules, thus removing planning uncertainties and the need for costly, time-consuming, case-by-case (or site-by-site) adversary proceedings except in borderline cases. By using this technology, public agencies can open up the decision-making process and can account for public values in formulating and implementing actions and policies for the public good. Such policies can easily be changed in response to changing value systems. This last point is very important as can be shown in the example presented here. One could easily imagine circumstances under which drastic oil shortages would produce shifts in public values about oil refinery construction decisions.

This chapter discusses the technique of multi-attribute utility measurement and illustrates how a version of it called SMART (Simple Multi-Attribute Rating Technique) can be used in coastal zone decision making. The illustration is in the form of a case study that takes place in the California coastal zone. The chapter then concludes with a suggestion of how a SMART-based decision-making process could be used in typical coastal zone decision-making situations. The ideas presented here are based on Gardiner,[1] and are related to those contained in Gardiner and Edwards.[2]

Coastal Zone Decision-Making

In the view of decision analysts, all decision making is seen to take place in an atmosphere of conflict over what is at stake and what the odds are (i.e., utilities and probabilities). To say that current coastal zone decision making fails to take these categories of conflict into account would simply not be true. To anyone who has observed public officials and public agencies in the process of making such decisions, the bulk of the action seems to center on figuring out "who thinks what will happen to whom," how "whom" is going to like it, and how much clout the respective "whoms" have. The debates over these points tend to take place in an adversary atmosphere that, not surprisingly, produces polarization of views and in the process promotes rather than reduces conflict. Each party to a decision appears before the public decision makers loaded with biased facts and dubious, overstated arguments aimed at persuading the decision makers of the wisdom of their position and the folly of opposing viewpoints. Advocates encourage discussion of any and all "facts" that support their case and try to enlarge their saliency while trying to ignore or minimize the saliency of "facts" that point in different directions. There is confusion about what information set is required to make the decision, and there is no orderly progression linking facts to

[1] Peter C. Gardiner, "The Application of Decision Technology and Monte Carlo Simulation to Multiple Objective Public Policy Decision Making: A Case Study in California Coastal Zone Management" (Ph.D. dissertation, Center for Urban Affairs, University of Southern California, Los Angeles, Ca., 1974.)

[2] Peter C. Gardiner and Ward Edwards, "Public Values: Multi-Attribute Utility Measurement for Social Decision Making," in *Human Judgment and Decision Processes: Formal and Mathematical Approaches,* Steven Schwartz and Martin Kaplan, eds. (New York: Academic Press, 1975.)

values, and values to a decision. The decision, when it comes, is often a compromise aimed at satisfying a key configuration of constituents. It has often been stated that public decisions are usually achieved by developing a course of action (or a policy) that can gather 50 percent of the decision-making vote plus one. But the process is largely intuitive. Decision makers often have no formal way of organizing and making sense of the bewildering array of information presented and consequently have little choice but to select pieces of information that seem relevant as best they can, combine them with a large dose of intuition and political savvy, and make a seat-of-the-pants decision.

It does seem reasonable, however, to make available to decision makers a more formal way of making coastal zone decisions for the public good. And any such way of formalizing the process should incorporate public values. The recent rash of "sunshine" legislation designed to open up public decision-making processes would seem to add some urgency to this effort. Since decision analysis has focused for some years on decision-making processes, it seems reasonable to turn to the set of tools that decision analysis research has provided to look for help.

Measuring Public Values

One reason for measuring public values is so that they can be used to help evaluate coastal zone actions or policies under consideration. And evaluation means literally that—the attaching of values to the impacts or effects of policies. These values are explicit, numerical answers to the question: "Is that particular effect good or bad, and how good or how bad?" Another way of putting the question to be answered in evaluating effects is: "Suppose I knew for certain that the policy under consideration would lead to a particular set of effects. How attractive would the policy be to me, or to someone, or to some collection of people?"

Note that *effects* of policies, not policies themselves, are evaluated. We could think of values as being attached to policies themselves. That is appropriate since the whole purpose of any evaluation effort is to select a policy for implementation. Yet, it is really the effects of the policy, not the policy itself, that are being evaluated.

Evaluation begins when decision makers try to identify and then discover or measure *relevant* effects of proposed policies. Yet, the mere collection and categorizing of effects is not an evaluation.

Decision makers need to know if a policy is a good idea. "Should a refinery be built?" "Where should refineries be built if we decide we need to build them?" A collection of numbers cannot tell a decision maker the answers to any of these kinds of questions— though it can offer hundreds of numbers that bear on them in one way or another.

Effects can be evaluated in many ways. Perhaps the most commonly used way is to assign economic proxies to effects. The phrase "That's so cheap it can't be any good" is true enough often enough to illustrate the phenomenon, and its frequent falsity illustrates how unsatisfactory the procedure is. Price depends more on the relation between supply and demand than on value. Also, there are some effects for which dollar proxies cannot be developed. Economists frequently assign dollar values to human lives lost in traffic accidents due to increased congestion following construction in the coastal zone; but if it were your life, would you sell it for that amount?

A more reasonable procedure is simply to consider the effect directly and make an intuitive value judgment. We all do this every day, and so do public decision makers. Yet, this too is an extremely primitive way of making value judgments.

Most effects have value for a number of different reasons. In our example, there is a growing recognition that there is a national energy shortage. New oil refinery construction may help combat the shortage, produce local jobs during construction and later during actual operations, and add to the local tax base. New oil refinery construction may also endanger the coastal zone with increased oil spills, increased air pollution, destruction of wildlife and marine life and so on as well as transforming a beautiful tourist area into an eyesore. All these considerations, and many others, may enter into a decision about whether or not to allow the construction of a new oil refinery at some shore location. Clearly, instances involving a multiplicity of effects present a multiplicity of problems. Who determines what effects are relevant to the evaluation and how relevant each is? How is that set of judgments made and used? Finally, what method is to be used to translate all this input information into a policy evaluation?

Multi-Attribute Utility Measurement (MAUM)

An explicit technology exists to answer some of these questions: Multi-Attribute Utility Measurement (MAUM). Expositions of various versions of it have been presented by Raiffa,[3] Keeney,[4] Edwards,[5] Gardiner and Edwards,[6] and others. Von Winterfeldt[7] presents an overview, integration and an evaluation of utility theory for decision analysis which discusses many utility models.

The essence of multi-attribute utility measurement, in any of its versions, is the collection of two kinds of value judgments from those whose values are considered important. The first value judgment is the relative importance of the various effects being measured. Each individual supplying value information assigns relative importance weights for each policy effect. The second value judgment involves measuring how preferences change with increasing or decreasing amounts of each effect (i.e., varying levels of policy performance on each of the value dimensions). The value of the overall policy is found by the use of an appropriate aggregation rule to combine the value "components." The aggregate value represents overall policy value. This method of evaluation can be used for each viewpoint considered important to the policy evaluation.

The results of such an evaluation process answer questions such as "Where are the strengths and weaknesses of the policies being considered and who says so?" "Overall, is there agreement or disagreement as to the worth of the policy?" "If there is disagreement, where exactly does it lie and to what extent?" Being able to respond to such questions is the true mark of a public evaluation process.

The version of multi-attribute utility measurement used in this paper is called SMART (simple multi-attribute rating technique). This version was first published by Edwards[8] and Gardiner and Edwards,[9] and it is oriented toward easy communication and use in environments in which time is short and decision makers are multiple and busy. Still, as Von Winterfeldt argues, SMART produces results essentially the same as much more complicated procedures involving imaginary lotteries which are oriented toward mathematical sophistication and formal utility theory.

[3] H. Raiffa, "Preferences for Multi-attribute Alternatives" (Rand Corporation Memorandum RM-5078-DOT/RC, April 1969).

[4] R. L. Keeney, "Utility Functions for Multi-attribute Consequences," *Management Science* 18 (1972):276-87.

[5] Ward Edwards, "Social Utilities," *The Engineering Economist*, Summer Symposium Series, VI (1971).

[6] Gardiner and Edwards, "Public Values: Multi-Attribute Utility Measurement."

[7] D. Von Winterfeldt, "An Overview, Integration, and Evaluation of Utility Theory for Decision Analysis," SSRI Technical Report 75-9, Social Science Research Institute, University of Southern California, Los Angeles, Ca., 1975.

[8] Edwards, "Social Utilities."

[9] Gardiner and Edwards, "Public Values: Multi-Attribute Utility Measurement."

SMART consists of ten steps:

Step 1: Identify the person or organization whose values are to be maximized. If, as is often the case, several organizations have stakes and voices in the decision, they must all be identified. People who can speak for them must be identified and induced to cooperate.

Step 2: Identify the issue or issues (i.e., decisions) to which the values needed are relevant. The same objects or acts may have many different values, depending on context and purpose. In general, value is a function of the evaluator, the entity being evaluated, and the purpose for which the evaluation is being made. The third argument of that function is sometimes neglected.

Step 3: Identify the entities to be evaluated. Here the focus is on public policy selecting among alternative courses of action.

Step 4: Identify the relevant dimensions of value. The first three steps were more or less philosophical. The first answered the question: whose value? The second answered the question: value for what purpose? The third answered the question: value of what entities? With Step 4, we come to the first technical task: discover what effects are important to the evaluation of any proposed policy (one effect = one value dimension).

As Raiffa has noted,[10] goals ordinarily come in hierarchies. But it is often practical and useful to ignore their hierarchical structure and instead to specify a simple list of goals that seem important for the purpose at hand.

It is important not to be too expansive at this stage. The number of relevant dimensions of value should be kept down, for reasons that will be apparent shortly. This can often be done by restating and combining goals or by moving upward in a goal hierarchy. Even more important, it can be done by simply omitting the less important goals. There is no requirement that the list evolved in this step be complete, and there is much reason to hope that it won't be.

Step 5: Rank the dimensions in order of importance. This ranking job, like Step 4, can be performed either by an individual, by representatives of conflicting values acting separately, or by those representatives acting as a group.

Step 6: Rate dimensions in importance, preserving ratios. To do this, start by assigning the least important dimension an importance of 10. (Use 10 rather than 1 to permit subsequent judgments to be finely graded and nevertheless made in integers.) Now consider the next-least-important dimension. How much more important (if at all) is it than the least important? Assign it a number that reflects that ratio. Continue on up the list, checking each set of implied ratios as each new judgment is

---

[10] Raiffa, "Preferences for Multi-Attribute Alternatives."

made. Thus, if a dimension is assigned a weight of 20, while another is assigned a weight of 80, it means that the 20 dimension is one-fourth as important as the 80 dimension. And so on. By the time you get to the most important dimensions, there will be many checks to perform; typically, respondents will want to revise previous judgments to make them consistent with present ones. That's fine; they can do so. In this step, individual differences are likely to arise.

Step 7. Sum the importance weights, divide each by the sum, and multiply by 100. This is a purely computational step which converts importance weights into numbers that, mathematically, are rather like probabilities. The choice of a 0-to-100 scale is, of course, purely arbitrary.

At this step, the folly of including too many dimensions at Step 4 becomes glaringly apparent. If 100 points are to be distributed over a set of dimensions and some dimensions are very much more important than others, then the less important dimensions will have nontrivial weights only if there aren't too many of them. As a rule of thumb, 8 dimensions is plenty, and 15 is too many. Knowing this, you will want at Step 4 to discourage respondents from being too finely analytical; rather gross dimensions will be just right. Moreover, the list of dimensions may be revised later, and that revision, if it occurs, will typically consist of including more rather than fewer.

Step 8: Measure the value of each policy being considered for each dimension. The word "measure" is used rather loosely here. There are three classes of dimensions: purely subjective, partly subjective, and purely objective. The purely subjective dimensions are perhaps the easiest; you simply get an appropriate expert to estimate the position of a policy on that dimension on a 0-to-100 scale, where 0 is defined as the minimum plausible value on that dimension and 100 is defined as the maximum plausible value. Note "minimum and maximum plausible" rather than "minimum and maximum possible." The minimum plausible value often is not total absence of the dimension.

A partly subjective dimension is one in which the units of measurement are objective, but "measuring" how a policy performs on it involves subjective estimation.

A wholly objective dimension is one that can be measured rather objectively, in objective units, before the decision. For partly or wholly objective dimensions, it is necessary to have the estimators provide not only values for each entity to be evaluated, but also minimum and maximum plausible values, in the natural units of each dimension.

At this point, there is a difference of opinion among users of multi-attribute utility measurement. Some are content to draw a straight line connecting maximum plausible with minimum plausible values, and then to use this line as the source of transformed location measures.[11] Others advocate the development of dimension-by-dimension value curves.[12] Of various ways of obtaining such curves, the easiest way is simply to ask the respondent to draw graphs. The X-axis of each such graph represents the plausible range of performance for the effect under consideration measured on its natural scale. The Y-axis represents the ranges of values, or desirabilities, or utilities associated with the corresponding X measurement. To find the dimension-by-dimension value contributions of each effect of a proposed policy, simply use the value curves to translate each effect measured on its own natural scale into a value-scale reading for that measurement. At this point, the effects of policies being evaluated have all been translated from incommensurable, "apples and oranges" scales to common value scales.

In what sense, if any, are the value scales comparable? The question cannot be considered separately from the question of what "importance," as it was judged at Step 6, means. Formally, judgments at Step 6 should be designed so that when the output of Step 7 (or of Step 6, which differ only by a linear transformation) is multiplied by the output of Step 8, equal numerical distances between these products on different dimensions correspond to equal changes in desirability. For example, suppose entity A has a value scale location of 50 and entity B a value scale location of 10 for an effect on dimension X, while A has a value scale location of 70 and B a value scale location of 90 for an effect on dimension Y (only X and Y are relevant). Suppose further that dimension Y is twice as important as dimension X. Then A and B should be equivalent in value. (The relevant arithmetic is: for A, 50 + 2(70) = 190; for B, 10 + 2(90) = 190. Another way of writing the same arithmetic, which makes clearer what is meant by saying that equal numerical differences between these products on different dimensions correspond to equal changes in desirability, is (50 - 10) + 2(70 - 90) = 0. It is important that judges understand this concept as they perform both steps 6 and 8.

Step 9: Calculate values for policies. The equation is

$$V_i = \Sigma_j w_j V_{ij}$$

remembering that $\Sigma_j v_j = 100$. V_i is the aggregate value for the ith policy, w_j is the normalized importance weight of the jth dimension of value, and v_{ij} is the rescaled position of the ith policy on the jth dimension. Thus, w_j is

---

[11] Edwards. "Social Utilities."

[12] H. Raiffa, *Decision Analysis: Introductory Lectures on Choices Under Uncertainty* (Reading, Mass.: Addison Wesley, 1968).

the output of Step 7 and u_{ij} is the output of Step 8. The equation, of course, is nothing more than the formula for a weighted average.

Step 10: Decide. If a single policy is to be chosen; the rule is simple: maximize V_i. If a subset of i is to be chosen, then the subset for which $\Sigma i V_i$ is maximum is best.

A special case arises when one of the policy effects, such as cost, is subject to an upper bound—that is, there is a budget constraint. In that case, Steps 4 through 10 should be done ignoring the constrained dimension. The ratios V_i/C_i should be chosen in decreasing order of that ratio until the budget constraint is used up. (More complicated arithmetic is needed if programs are interdependent or if this rule does not come very close to exhausting exactly the budget constraint.) In the absence of budget constraints, cost is just another dimension of value, to be treated on the same footing as a-1 other effects of value.

The multi-attribute utility approach can easily be adapted to cases in which there are minimum or maximum acceptable values on a given dimension of value, by simply excluding policy alternatives that lead to effects that transgress these limits.

It should be noted that practically every technical step in the preceding list has alternatives. For example, Keeney has proposed use of a multiplicative rather than an additive aggregation rule.[13] Certain applications have combined multiplication and addition. The methods suggested above for obtaining location measures and importance weights have alternatives; the most common is the direct assignment of importance weights on a 0-to-100 scale. (This procedure is considered by some to be inferior to the one described above, but there is reason to doubt that it makes much practical difference in most cases.)

Because the emphasis here is on simplicity and on rating rather than on more complicated elicitation methods, the above technique is called a Simple Multi-Attribute Rating Technique (SMART).

One final technical point involves value independence. Either the additive or the multiplicative version of the aggregation rule assumes value independence. Roughly, that means that the extent of your preference for location a_2 over location a_1 of dimension A is unaffected by the position of the policy being evaluated on dimensions B, C, D, ... Value independence is a strong assumption, not easily satisfied. Fortunately, in the presence of even modest amounts of measurement error, quite substantial amounts of deviation from value independence will make little difference to the ultimate number V_i, and even less to the rank ordering of the V_i values. Moreover, if the value curves are all

---

[13] R. L. Keeney, "Multiplicative Utility Functions," Technical Report No. 70, Operations Research Center, MIT, Boston, 1972b.

monotonic, an additive approximation (e.g., SMART) will almost always work well.

Nothing in the preceding discussion ensures that different respondents will come up with similar numbers—and such agreements are indeed rare.

One might expect that the magnitude of interpersonal disagreement would make this technology of questionable value. Not so. Consider the following application in a case study of coastal zone management.

Selecting the Case Study

A review of the national population in the 1950, 1960, and 1970 census shows that there has been and continues to be population growth and concentration in the nation's coastal zone (by which we mean coastal counties). While not uniformly distributed throughout the coastal zone, urban concentrations continue to develop and expand within it. The national coastal county population, as a percentage of the total national population has risen from 37 percent in 1950 to 41 percent in 1970. Coastal zone standard metropolitan statistical areas (SMSA's) contained 45 percent of the population of all SMSA's in the nation.[14] The concentration of growth in the coastal zone has produced drastic changes and fierce competition for the use of its scarce resources. The result is problems of ecology, conservation, development, land use, transportation, public access, recreation, public utilities, maintenance of ocean mineral and fish and game resources, education and science, etc. The growing complexity of the problems and issues involved in coastal zone management calls for a level of sophistication that is straining the capacities of local, urban, and county governments. The consequences of current evaluations and decision making can be seen all too clearly in pollution, urban sprawl, esthetic blight, disappearing resources, and so forth. In fact, sixty of the nation's most distinguished academic and professional individuals directly involved in coastal zone management issues have concluded that there is a pressing need to develop a systematic way to study and express the value systems that affect management practices in coastal zone activities.

In 1972, the state of California found itself thrust into a position of leadership in developing policies and programs for the planning and management of its coastal zone. This is partially because both the state and coastal county populations doubled between 1950 and 1960. The state population increased from 10 million to almost 20 million, while the

---

[14] These figures result from research on coastal zone management done in 1972 by the author while with the USC Sea Grant Program.

coastal county population rose from approximately 9 million to about 17 million, or 85 percent of the state's population.

California has more than 1,100 miles of coastline, including wide sandy beaches, scenic bluff areas, and beautiful rocky headlands that jut into the sea. It is estimated that in the last two hundred years, the length of California beaches and shoreline available to the public has shrunk from 1,062 miles to approximately 200 miles.

Prior to 1972, two hundred separate entities—city, county, state, and federal governments, agencies, and commissions—regulated the California coast. The citizens of California, in reviewing the performances of these two hundred entities, were apparently dissatisfied and, in a voter-sponsored initiative during the general election of 1972, the voters approved legislation placing coastal zone planning and management under one state commission and six regional commissions. In passing the *Coastal Zone Conservation Act* by 55 percent of the vote, the voters established decision makers with ultimate authority (other than appeal to the courts) to preserve, protect, restore, and enhance the environment and ecology of the state's coastal zone.

The coastal zone is defined in the Act as the area between the seaward limits of state jurisdiction and 1,000 yards landward from the mean high tide line. The Act requires two categories of activity from the state and regional commissions. First, any plan for development within the coastal zone must be approved by the appropriate regional commission before it can be carried out. Disapprovals can be appealed to the state commission and then to the courts if necessary. (Development permits are similar to other types of building permits and authorize only the specific activities named.) The second responsibility of the commissions under the Act is to prepare and submit to the state legislature a California coastal zone conservation plan. The plan must include a "precise, comprehensive definition of the public interest in the coastal zone" and a land-use element, a transportation element, a public access element, a population density element, a recreation element, and so on.

The South Coast Regional Commission (Region V), comprising Los Angeles and Orange counties, is one of the six regional commissions. Los Angeles county is heavily urbanized, and in 1970 it contained 35 percent of the total state population and 41 percent of the state's coastal county population. Los Angeles County includes the coastal cities of Long Beach, Redondo Beach, Hermosa Beach, Manhattan Beach, Los Angeles (Venice and the harbor area), Santa Monica, and unincorporated county areas such as Marina del Rey. These cities and areas all contain portions of the coastal zone that are under the control of the Region V Commission.

Approximately $1 billion worth of development was authorized in the first year of the commission's activities, and more than 1,800 permits were acted upon. A backlog as high as 600 permit requests awaiting action has existed. The evaluation and decision-making tasks that confront the Region V Commission members are important, far-reaching, difficult, and controversial.

The Case Study[15]

The Coastal Zone Conservation Act has charged the commissioners of the Region V Commission with evaluating the worth of each development request submitted and then either approving or disapproving. A major problem results since the Act does not specify just how this very large and difficult evaluation and decision-making task is to be accomplished. The permit requests consist of information on many different importance dimensions that are specified (at the abstract, conceptual level) by the Act. Decisions taken on permits prior to the development of a master plan are to be consistent with the eventual planning output.

Although the Act specifies that certain attributes should be considered in making evaluations, it fails to specify just how they are supposed to enter into the evaluation process. Nor does the Act specify how the Commissioners are to balance the conflicting interests affected by their decisions. In effect, the Act implies that individual commissioners assigned to the commission will represent the interests of all affected parties with respect to the coastal zone in Region V. How this is to be accomplished in practice is left unspecified. In practice, attempts to include the preferences and value judgments of interested groups and individuals occur when the commission holds public advocacy hearings on permit requests. Under these procedures, opposing interest groups express their values and viewpoints as conclusions—often based on inconsistent sets of asserted facts or no facts at all—in the form of verbal and written presentations at the open hearings.

Fourteen individuals involved in coastal zone planning and decision-making agreed to participate in this study. Included were two of the current coastal commissioners for Region V, a number of active conservationists, and one major coastal zone developer. The purpose of this study was to test the consequences of using multi-attribute utility

[15] For a complete discussion of this case study, see Gardiner, "The Application of Decision Technology and Monte Carlo Simulation to Multiple Objective Public Policy Decision Making: A Case Study in California Coastal Zone Management."

measurement processes by having participants in or people close to the regulatory process with differing views, rather than the usual college sophomores, make both individual and group evaluations of various proposals for development in a section of the California coastal zone. Evaluations were made both intuitively and by constructing multi-attribute utility measurement models.

To provide a common basis for making evaluations, a sample of fifteen hypothetical but realistic permit requests for development were "invented." The types of permits were limited to those for development of single-family dwellings, duplex, triplex, or multi- family dwellings (owned or for renting). Dwelling unit development (leading to increased population density) is a major area of debate in current coastal zone decision making. Most permit applications submitted to the Region V Commission thus far fall into this class. Moreover, permits granted in this class will probably generate further permit requests. Housing development tends to bring about the need for other development in the coastal zone such as in public works, recreation, transportation, and so on. The permit applications provided eight items of information about the proposed development that formed the information base on which subjects were asked to make their evaluations. These eight items were abstracted from actual staff reports currently submitted to the Region V coastal commissioners as a basis for their evaluations and decision making on current permit applications. The commissioners' staff reports do have some additional information such as the name of the applicant, and so on, but the following items are crucial for evaluation.

1. *Size of development*: Measured in the number of square feet of the coastal zone taken up by the development

2. *Distance from the mean high tide line*: The location of the nearest edge of the development from the mean high tide line measured in feet

3. *Density of the proposed development*: The number of dwelling units per acre for the development

4. *On-site parking facilities*: The percentage of cars brought in by the development that are provided parking space as part of the development on-site

5. *Building height*: The height of the development in feet (17.5 feet per story)

6. *Unit rental*: Measured as the dollar rental per month (average) for the development; if the development is owner-occupied

and no rent is paid, an equivalent to rent is computed by taking the normal monthly mortgage payment

7. *Conformity with land use in the vicinity*: The density of the development relative to the average density of adjacent residential lots; measured on a 5-point scale from much less dense to much more dense

8. *Esthetics of the development*: Measured on a scale from poor to excellent

Each of the invented permits was constructed to report a level of performance for each item. They were as realistic as possible and represented a wide variety of possible permits.

Each subject answered seven questionnaires. In general, the participants had five days to work on each of the questionnaires. Throughout, the procedures of the Delphi technique were used.[16] In the process of responding to the seven questionnaires, each subject (1) categorized him/herself on an 11-point continuum that ranged from very conservationist-oriented to very development oriented, (2) evaluated intuitively (wholistically) fifteen sample development permit requests by rating their overall merit on a 0-to-100 point worth scale, (3) followed the steps of SMART outlined previously and in so doing constructed individual and group value models,[17] and (4) reevaluated the same fifteen sample permit requests intuitively a second time. Subjects did not know that the second batch of permits was a repetition of the first.

The location of the proposed developments was Venice, California, which is geographically part of the city of Los Angeles located between Santa Monica and Marina del Rey. Venice has a diverse population and has been called a microcosm—a little world epitomizing a larger one. In many ways, Venice presents in one small area the most controversial issues associated with coastal zone decision making.

After the initial questionnaire round in which the subjects categorized themselves according to their views about coastal zone development, the fourteen individuals were divided into two groups. The eight more conservationist-minded subjects were called Group 1 and the other six

---

[16] The use of this technique has become fairly common. It was developed by Norman Kalkey and Olaf Helmer in the 1960s. See N. C. Dalkey, *Delphi* (Santa Monica: Rand Corp., 1968).

[17] The evaluation and decision making in this study are assumed to be riskless. Decisions involving permit requests, by the nature of the permits themselves, suggest that the consequences of approval or disapproval are known with certainty. The developer states on his permit what he intends to do if the permit is approved and is thereby constrained if approval is granted. If the request is disapproved, there will be no development. Outcomes are known with certainty, and this is the requirement for riskless decision making.

subjects (whose views, by self-report, range from moderate to strongly pro-development) were called Group 2.

In both the intuitive evaluation and multi-attribute utility measurement tasks, the subjects reported no major difficulty in completing the questionnaires. An example of one participant's value curves and importance weights is shown in Figure 1. The abscissae represent the natural dimension (effects) ranges, and the ordinates represent worth ranging from zero to 100 worth points.

To develop group intuitive ratings and group value models, each individual in a group was given, through feedback, the opportunity of seeing his group's initial responses on a given task (intuitive ratings, importance weights, etc.). These data were fed back in the form of group means. Individual responses within a group were averaged to form group responses.

Results of the Case Study

From the point of view of this paper, the interesting question to be answered is "What is the effect of using a group's value model versus a group's intuitive evaluation?" To answer this question, a two-way analysis of variance was conducted of permit worths by group by permit; the results are shown in Table 1. These results indicate that the two groups initially (i.e., by wholistic intuitive evaluations) represented differing viewpoints (i.e., were drawn from differing populations) although the differences were not dramatic. Substantial percentages of variance were accounted for both by group main effects and by permit-group interactions for the first-test wholistic evaluations. (Results for the retest were similar and are not presented here.) Both findings indicate differing viewpoints between the two groups. The main effect could be caused, however, by a constant evaluation bias alone. The key indication of differing viewpoints is the interaction term. Notice that use of each group's value model evaluations instead of their intuitive evaluations causes the percent of variance accounted for by the interaction to drop from 12 percent to 2 percent. Figure 2 shows this difference dramatically.

In other words, *use of SMART has turned modest disagreement into substantial agreement.*

Why? Let me suggest a plausible answer. When making wholistic evaluations, those with strong points of view tend to concentrate on those aspects of the entities being evaluated that most strongly engage their biases. The SMART procedure does not permit this; it separates judging the relative importance of an effect from judging the value of whatever

Figure 1

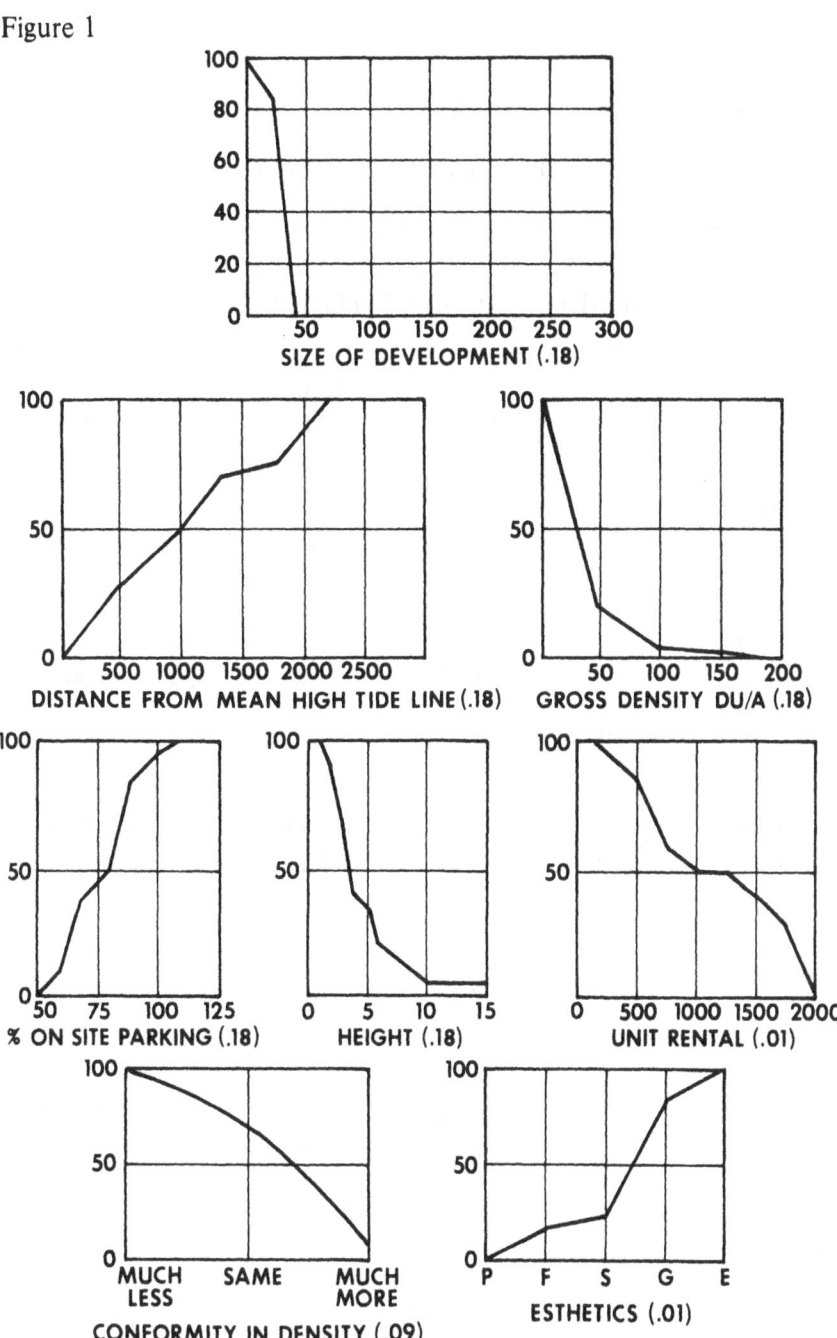

An Example of Value Curves and Importance Weights (in parentheses) for Permit Items—for Group 1 Subject

Table 1

Analysis of Variance Summary

TWO-WAY ANALYSIS OF VARIANCE (PERMIT X GROUP) FOR WHOLISTIC EVALUATION PERMIT WORTH

Source	d.f.	MS.	Error d.f.	F	P
Main Effect					
Permits	14	4274.6668	180	9.91	0.0005
Group	1	13675.2366	180	31.70	0.0005
Interaction					
Permit/group	14	1536.8741	180	3.56	0.0005
Within Cells	180	431.3517			

TWO-WAY ANALYSIS OF VARIANCE (PERMIT X GROUP) FOR SMART EVALUATION PERMIT WORTH

Source	d.f.	MS.	Error d.f.	F	P
Main Effect					
Permits	14	1853.0008	180	17.47	0.0005
Groups	1	2128.5145	180	20.06	0.0005
Interaction					
Permit/group	14	77.6310	180	0.73	0.741
Within Cells	180	106.0942			

Table 1 (cont.)

PERCENT OF VARIANCE ACCOUNTED FOR

Rating	Source	Percent in Sample	Percent in Population (estimate)
Wholistic	Permit	0.34	0.31
SMART	Permit	0.53	0.49
Wholistic	Group	0.08	0.08
SMART	Group	0.04	0.04
Wholistic	Interaction	0.12	
SMART	Interaction	0.02	
Wholistic	Within Cells	0.45	
SMART	Within Cells	0.39	

182

Figure 2

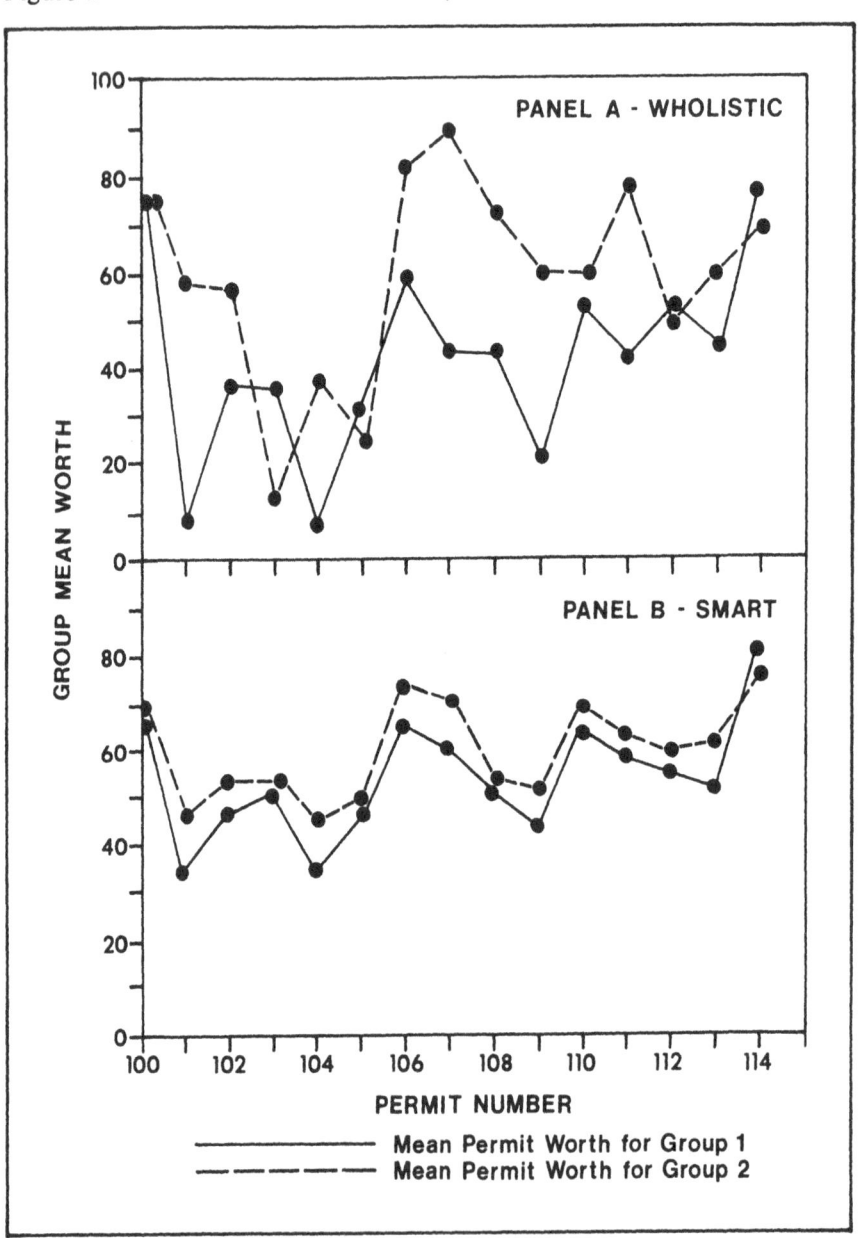

SMART-Fostered Agreement

amount of it is produced by a particular policy. These applications varied on eight dimensions that are relevant to the environmentalist-versus-builder arguments. While these two views may cause different thoughts about how good a particular level of some effect may be, evaluation of other effects will be more or less independent of viewpoint. Agreement about those other effects tends to reduce the impact of disagreement on controversial effects. That is, multi-attribute utility measurement procedures do not provide an opportunity for any one or two effects to become so salient that they emphasize existing sources of conflict and disagreement. Multi-attribute utility measurement cannot and should not eliminate all disagreement however; such conflicts are genuine, and any value measurement procedure should respect and so reflect them. Still, in spite of disagreement, social decisions must be made. How?

There are two kinds of disagreements. Disagreements in measuring the effect itself at Step 8 are essentially like disagreements among different thermometers measuring the same temperature. If they are not too large, there is little compunction about taking an average. If they are large, then we are likely to suspect that some of the thermometers are not working properly and to discard their readings. In general, judgmentally determined location measures of the effects of policies should reflect expertise, and typically one might expect different effects to require different kinds of expertise and therefore different experts. In some practical contexts, the problem of measurement disagreement at Step 8 can be avoided entirely by the simple expedient of asking only the best available expert for each dimension to make measurement judgments about that dimension.

Disagreement at Steps 5, 6, and value curve construction in Step 8 are another matter. These seem to be the essence of conflicting values, and it is important to respect them as much as possible. For that reason, the judges who perform them should be well-chosen representatives of the public. Considerable discussion, persuasion, and information exchange should be used in an attempt to reduce disagreements as much as possible. At the least, this process offers a clear set of the rules of debate and an orderly way to proceed from information and data to values and to decisions.

Even this will seldom reduce disagreements to zero, however. For some organizations, it becomes necessary to invoke former President Truman's desk sign, "the buck stops here." One function of executives, bosses, or public decision makers is to resolve disagreements among the public. This can be done in various ways: by substituting their judgment for the public's, by picking one segment of the public as "right" and rejecting the others, or, in the weighted-averaging spirit of multi-attribute utility measurement, by assigning a weight to each of the disagreeing publics and then calculating weighted-average importance weights.

If there is no individual or group decision maker to resolve disagreement, the evaluation can be carried out separately for each of the disagreeing individuals or groups, hoping that the disagreements are small enough to have little or no action implications. And if that hope is not fulfilled, there are no suggestions to offer beyond the familiar political processes by means of which society functions in spite of conflicting interests. SMART offers technology, not miracles!

A Technology for Coastal Zone Decision-Making

Let me conclude with a suggestion of how SMART-based decision-making processes could be institutionalized in coastal zone decision-making. The goal is to have coastal zone decision-making agencies carry out their decision-making tasks by fully exploiting SMART or some similar value measurement technique.

Public statutes would define, at least to some degree, the appropriate effects (value dimensions) as they do now. They might also, but probably should not, specify the limits on the importance weights attached to these dimensions. They might, and perhaps should, specify boundaries beyond which no value could go in the undesirable direction. At this point, the statutes have done their work and a coastal zone regulatory agency takes over the process.

The regulatory agency would have four main functions: to specify measurement methods for each effect (with value functions or other methods for making the necessary transformations at Step 8); acceptable levels of aggregated value, and perhaps also lower bounds not specified by statute on specific effects; and to hear appeals. Perhaps two bounds on acceptable levels of aggregated value would be appropriate. Requests falling above the higher bound would be accepted automatically; requests falling below the lower bound would be rejected automatically; requests falling in between would be examined in old-style hearings. Presumably the public regulatory agency would also have to hear appeals from the automatic decisions, perhaps with the provision that the appellant must bear the cost of the hearing if the appeal is rejected.

The regulatory agency could afford to spend enormous amounts of time and effort on its first two functions—specification of measurement methods and of importance weights. Value considerations, political considerations, views of competing constituencies and advocates, the arts of logrolling and compromise—all would come into play. Public hearings would be held, with elaborate and extensive debate and full airing of all relevant issues and points of view.

The regulatory agency would have further responsibilities in dealing with measurement methods for wholly or partly subjective value dimensions. Since such measurements must be judgments, the regulatory agency must make sure that the judgments are impartial and fair. This could be done by having staff members make them, or by offering the planner a list of agency-approved impartial experts, or by mediating among or selecting from the conflicting views of experts selected by those with stakes in the decision, or by some combination of these methods. The first two of these approaches are probably the most desirable, but the third or fourth may be inevitable.

The costs of prolonged and intensive study of measurement methods and of importance weights could be borne because they would recur infrequently. Once agreed-on measurement methods and importance weights had been hammered out, most case-by-case decisions would be made automatically by means of them. Only in response to changed political and social circumstances or changed technology would reconsideration of the agreed on measurement methods and importance weights be necessary—and even such reconsiderations would be likely to be partial rather than complete. They would, of course, occur; times do change, public tastes and values change, and technologies change. Those seeking appropriate elective offices could campaign for such changes; an election platform consisting in part of a list of numerical importance weights would be a refreshing novelty.

The decision rules would, of course, be public knowledge. That fact probably would be the most cost-saving aspect of this whole approach. Planners would not waste time and money preparing plans that they could easily calculate to be unacceptable. Instead, they would prepare acceptable plans from the outset. Once a plan had been prepared and submitted to the regulatory agency, its evaluation would consist of little more than a check that the planner's measurements and arithmetic had been done correctly. Delay from submission to approval need be no more than a few days.

Changes in the decision rules can be and should be as explicit and public as the rules themselves. Such explicitness would permit both regulators and those regulated to know just exactly what current regulatory policies are, and, if they have changed, how and how much. Such knowledge would greatly facilitate both enlightened citizen participation in deciding on policy changes and swift, precise adaptation of those regulated to such changes once they have taken effect.

In short, multi-attribute utility measurement allows value conflicts bearing on social decisions to be fought out and resolved at the level of decision rules rather than at the level of individual decisions. Such decision rules, once specified, define and thus remove nearly all ambiguity from regulatory policy without impairing society's freedom to modify policies in response to changing conditions. Possible savings in financial and social costs, delays, frustrations, and so on, are incalculable, but cost reduction in dollars alone could be enormous.

The idea of resolving value conflicts at the level of decision rules rather than at the level of individual decisions has the potential of revolutionary impact on public decision making. Any revolutionary idea is bound to be full of unexpected consequences, booby-traps, and surprises. For a while, therefore, the wise innovator would want to run old and new systems in parallel, comparing performance of the two and building up experience with the new system. The mechanism suggested above of defining an upper and a lower bound, with automatic acceptance above the upper bound, automatic rejection below the lower one, and hearings in between, provides a convenient administrative device for operation of such parallel procedures. Initially, the upper bound could be very high and the lower bound very low so that most cases fall in between and are handled by the traditional hearing mechanism. One candidate number for the lower bound, at least initially, is the value of the do-nothing (i.e., status quo) alternative, for obvious reasons. If what the applicant wants is not clearly better than the status quo, why does he deserve a hearing? As experience and confidence in the multi-attribute utility measurement system develop, the two bounds can be moved toward each other, so that more and more cases are handled automatically rather than by means of hearings. This need not work any hardship on any rejected applicant; he can always appear, accepting the delays, costs, and risk of losing implicit in the hearing process rather than the cost of upgrading his plan. And the regulatory agency, by moving the boundaries, can in effect control its case load, thus gradually shortening what are, under current procedures, frequently inordinate delays.

CHAPTER 9
THE VALUE OF THE OCEAN TO THE UNITED STATES: HISTORICAL ANTECEDENTS AND PRELIMINARY MEASUREMENTS

GIULIO PONTECORVO[*]

Historical Antecedents[1]

The utility of economic theory in analyzing the economic potential of ocean-based activity has been limited by two complementary difficulties: (1) the need to bring economic analysis of ocean affairs into the mainstream of economic theory and policy; and (2) the need for an analytical framework which contributes to the ordering of research and facilitates creation of research priorities in policy studies.[2] A partial solution to the first problem is to move from a micro to a macro level and to go beyond investigation of limited aspects of the subject toward a more general, theoretical framework. I do not advocate elimination of micro studies of specific areas; however, the place of these studies can best be defined by the larger analytical framework of economic theory and methodology currently used in policy analysis. Such a macro-economic framework would encompass partial equilibrium studies, and would also contribute to improving the methodology employed. This chapter contains a report on the historical background that lies behind the development of a project that contributes to the creation of the desired macro framework and additionally creates an order of research and policy priorities. In undertaking the project, our objective was simple: to measure the value of the oceans to the United States economy and to measure the output of the various components of ocean-based activities. It is necessary that such

[*] Giulio Pontecorvo is Professor in the Graduate School of Business at Columbia University. This paper reports on a joint project of the Columbia University School of Business and the United States Department of Commerce to create an ocean account within the framework of the National Income Accounting System. Professor Maurice Wilkinson is the co-principal investigator on the project.

[1] See Giulio Pontecorvo, Maurice Wilkinson, Ronald Anderson, and Michael Holdowsky, "Contribution of the Ocean Sector to the United States Economy," *Science*, 36(May 30, 1980): 1000-1006; and Working Paper 243A, Graduate School of Business, Columbia University, New York, N.Y., 1972 (unpublished).

[2] A. Scott, "Development of Economic Theory on Fisheries Regulation," *Journal of the Fisheries Research Board of Canada* 36(July 1979):725-741.

a set of numbers be consistent with the way we measure the rest of the economy. Thus, we set about creating an "ocean sector" within the existing analytical framework of National Income Accounts. It is the first spatial sector within that framework.

The new account makes economics more useful in the analysis of ocean activity. It opens up the possibility of model-building to permit measurement of the effects of ocean policy decisions on the entire economy, and vice versa. Since it establishes a set of relative values for the component activities, it also facilitates decisions about the relative importance of alternative research possibilities. However, before turning to the ocean account itself, it is desirable to review selected elements of the literature on utilization of ocean resources. This literature lies behind the development of the ocean account, and a review will serve to point up the methodological and analytical differences in the various approaches to the valuation of ocean activity and policy.[3] In some instances, for example in the so-called "Nathan Report,"[4] such literature is directly antecedent to the concept of an ocean account. In others, especially the work of Cooper and Bergsten, discussed below, it stresses a number of issues that transcend the concept of the ocean account but that indicate the importance of and the difficulties that surround the welfare economic analysis that underlies the ocean account. After evaluation of the background literature and the underlying welfare issues, this chapter will deal with the more general implications of the results of calculating the value of output from the oceans.

Literature: The State of the Art

Welfare Economics

Elsewhere, a report on the ocean-income project has spelled out the theoretical assumptions necessary for using actual market prices to establish the value of commodities.[5] The nature of the link between the

[3] Ross Eckert has recently provided an extended economic analysis of a wide range of ocean problems. This work is not part of the antecedent literature to the ocean accounting program. Further, Eckert's work requires greater consideration and evaluation than could be provided here. See Ross D. Eckert, *The Enclosure of Ocean Resources: Economics and the Law of the Sea* (Stanford, Ca.: Hoover Institution Press, Stanford University, 1979); and G. Pontecorvo's review of this work in the *Journal of Economic Literature* XVIII: 1618-19, December 1980.

[4] U.S. Congress, Committee on Commerce, *The Economic Value of Ocean Resources to the United States*, 94th sess., 1974.

[5] Giulio Pontecorvo and Maurice Wilkinson, "From Cornucopia to Scarcity: The Current Status of Ocean Resource Use," *Journal of Ocean Development and International Law*

actually observed market price of a thing and its theoretically optimum price is crucial to the understanding of the meaning of national income. Markets that are subject to imperfections (e.g., monopoly elements, externalities, and public goods) tend to produce an observed price vector that differs significantly from the prices that are theoretically socially optimum for the commodities traded. Commodities that result from ocean economic activity are, in large measure, traded in imperfect markets and are subject to externalities; many involve public goods. Thus, much of the output from ocean resources is priced in a way that does not maximize economic well being; indeed, there is a substantial deviation of observed market price from the theoretically optimum price. Given the presence of various kinds of significant imperfections, the use of actual market prices to value commodities produced in the oceans has serious theoretical shortcomings.

On the other hand, the theory necessary to correct price from what is observed to what is optimal is currently fragmentary and is only in the early stages of development. As the required theoretical propositions necessary for the correction of market prices are derived, it may be possible to devise procedures by which to modify actual market prices by taxes and other policies so that public commodities and goods exchanged in these markets are more correctly valued.[6] Unfortunately, we are not now in a position to create, either theoretically or empirically, the appropriate price vector for the ocean sector. We must, therefore, work with the current state of the art of measurement of the economic value of ocean economic activity; that is, we must utilize the existing set of observed prices while keeping this underlying problem in welfare economics clearly in mind.[7] I will now turn to a consideration of the important antecedent studies on ocean resource use. My discussion does not exhaust the relevant literature, some of which is unpublished; however, the studies considered do allow me to spell out the methodological issues involved.

5(1978):383-395; and Pontecorvo and Wilkinson, et al., "Contributions of the Ocean Sector to the United States Economy."

[6] For an exposition of how well or how poorly deviations from optimal price can be corrected see Giulio Pontecorvo and John Donaldson, "Canada and the United States: Reconciliation of Interests in Atlantic Fisheries," *Journal of Ocean Development and International Law*. Note the discussion of the cost of information required to optimize a multi-species fishery and also a discussion of the limits of the output from management organization required to operate such a fishery.

[7] For a significantly different approach to the problem of welfare measurement using GNP, see Morris D. Morris, *Measuring the Condition of the World's Poor*, Pergamon Policy Series (New York: Pergamon Press, 1979). Also, for an exhaustive critique of current welfare measure, see Amartya Sen, "The Welfare Basis of Real Income Comparison: A Survey," *Journal of Economic Literature* 17(March 1979):1-45.

The Cooper and Bergsten Papers[8]

The determination of developing countries to work toward an improvement in the distribution of income throughout the world has been a substantive cause of the prolonged deliberations at the United Nations Conference on the Law of the Sea. The works of R. N. Cooper and C. F. Bergsten illustrate different policy solutions for this difficult international problem. These studies are important to the analysis of income accounting in that they build their arguments around the markets for ocean products, and, implicitly, they indicate the need for ordering priorities in terms of the relative value of the commodities derived from ocean resources.

Both Cooper and Bergsten adopt, as a working assumption, the general contention that the oceans are the common heritage of mankind. Cooper focuses his analysis on fisheries and the problem of maximizing the discounted present value of fish stocks by moving toward the efficient production of fishery resources. He draws upon the findings of the extensive literature on the economics of fisheries which begins with the hypothesis that because of the legal status of fish stocks (even under extended economic zones) as open access or common property resources, the fisheries of the world are overcapitalized. The overcapitalization and, therefore, the redundance of factor inputs (i.e., more labor and capital is employed than is necessary to catch the fish) means that the cost of inputs employed in fisheries tend to equal the total revenue generated. In these circumstances, if entry can be limited, a portion of the total revenue that is now paid to the labor and capital that is redundant in the productive process can be captured by a tax or license. By limiting entry, the industry may be made more efficient so that fewer inputs are used to capture the same quantity of fish. The income payments to these excess factors of production represent a rent; the income payments to the redundant factors are not necessary to sustain the given level of output. The amount of this rent may be quite large, and, in highly valued fisheries, it may reach fifty percent or more of the total value of the catch.

More precisely, within the framework of general equilibrium theory, this rent is the measure of the deviation of the market price from the

---

[8] R. N. Cooper, "The Oceans as a Source of Revenue," paper prepared for the MIT Workshop on the New International Economic Order, Cambridge, Mass., 1976; R. N. Cooper, "An Economist's View of the Ocean," in R. E. Osgood, ed., *Perspectives on Ocean Policy* (Washington, D.C.: Johns Hopkins University, 1974), pp. 145-165; and C. F. Bergsten, "Commodity Shortages and the Ocean," in Osgood, *Perspectives on Ocean Policy*, pp. 167-178.

price which would yield the appropriate (best welfare) allocation of resources in the fishery. In an attempt to solve this efficiency problem and to improve the international distribution of income, Cooper advocates a market-oriented approach. He solves the legal problem of common property (open access) by centralizing the ownership of world fisheries in the hands of an international authority. The creation of property rights in the world's fisheries could permit a normal market solution to the allocation (efficiency) and income distribution problem.

In today's world, with nationalism as a dominant force in international relations, one may have legitimate reservations about the likelihood of the nations of the world creating an international authority with substantive power over resources which coastal states now claim. In Cooper's proposed system, the designated international authority would auction off leases and licenses to fish to anyone who wanted to buy them. Then, it would tax the firms to capture any of the economic rent that had not been captured by the auction process. The proceeds from the auction and taxation would be dispersed through international agencies such as the United Nations Development Program and various regional development banks, to be used for research and development of ocean resources and, most importantly, to aid the development plans of the lesser developed countries (LDCs).

Cooper's plan assumes a relatively free market throughout the world. In these circumstances, and given certain additional assumptions about the relative cost and productivity of capital and labor in different countries, a lesser developed country would be better off buying the fish on the world market rather than engaging in fishing. Note that this argument rests on a second, and possibly very questionable, set of assumptions that the lesser developed countries do not have the technology, the required capital, or the productive labor to work these resources at as low a cost level as would the firms affiliated with more advanced countries. Cooper suggests that the LDCs would benefit from efficient production of the resources by the developed states since the tax proceeds from such activity will be, in part, supplied to them without any production effort on their part.[9]

If there are few fish stocks left, either unexploited or underexploited, the benefits from the creation of an international fisheries authority must lie in the ability of the authority to increase efficiency by imposing a market solution on world fisheries. The creation of a single ownership and

---

[9] See Giulio Pontecorvo and Maurice Wilkinson, "Comments on Cooper Plan for Obtaining Revenue from World Fisheries to be Used for Purposes of Economic Development (Income Redistribution) in the Third World," Columbia University, New York, N.Y., 1976 (unpublished).

management unit would permit capture of the rent from the increases in productivity that would result from the rationalization of world fisheries. These rents may then be used in ways that will assist humanity, specifically to improve the rate of economic growth in the third world. In addition, Cooper asserts that conservation practices would improve under the aegis of the authority.

A market solution like Cooper's requires that the classical assumptions about factor mobility, price flexibility, and free trade apply, to a reasonable extent, through the world; that is, the basic conditions for welfare optimization must be met. However, the available evidence casts doubt on these assumptions and also suggests that the national objectives of the LDCs include (in addition to Cooper's transfer payment from an international authority) employment considerations, control and operation of the technology utilized in fisheries (technology transfer), and national control over resources.[10]

In the developed states, the principal obstacle to any substantial economic rationalization of the fisheries has been the employment argument. This argument is even more critical throughout the third world. Most important, however, is the fact that the worldwide trend is toward more, not less, national control of resources. The creation in 1976 of a 200-mile economic zone by the United States (under pressure from domestic processors and fishing interests) is symbolic and representative of the position of the nations at the Law of the Sea (LOS) Conference. The final outcome of the conference cannot be foreseen, but one result that is already apparent is the extension of national jurisdiction in some form by the coastal states.

An international authority for fisheries is not even a subject for discussion at the LOS conference. It is because of the reasons enumerated (failure to meet the classical assumptions; the key role of the employment problem in both the short and the long run; the preference of the LDCs for technology transfer over transfer payments; and the pressure for nationalism which dominates all issues at the conference) that the traditional optimization model utilized by Cooper is an inappropriate framework for a general analysis of fishery problems.

Bergsten takes a different position from Cooper on both the availability of supply and the free-market approach. He believes, speaking primarily of sea bed resources, that there are no physical shortages of ocean resources—only contrived ones, that is, production restrictions in the form of quotas, political embargos, and so forth. He suggests that certain

---

[10] See Giulio Pontecorvo and Maurice Wilkinson, "An Economic Analysis of the International Transfer of Marine Technology," *Journal of Ocean Development and International Law* 2(1974):255-284.

LDCs (e.g., Chile, Peru, etc.), which are primary producers of basic minerals, and developed countries which are the largest users of minerals, have opposing interests in opening up the sea beds. While the latter have an interest in doing so, the former would rather protect their existing mines and conserve their share of the world market. More simply, the LDCs want to limit any new source of supply.

Bergsten believes that the distribution-equity issue between nations is more important than is emphasis on solutions that maximize efficiency. He suggests a political concensus at the international level. This derives from his assumption that the neo-classical economic welfare optimal conditions cannot be met in the case of ocean resource exploitation. For example, in the case of deep-sea mining, he suggests that the enormous capital outlays required are affordable only by a few firms in developed countries and not, by numerous economic agents.[11] The national objectives and supply conditions assumed by Bergsten lead to a producers' cartel and therefore obviate a classical market solution.

To Bergsten, a national solution is the only feasible one. If the United States were to go ahead with its own scheme of ocean activity, it would serve as a spearhead to dissolve the present international stalemate. Such action, presumably, would serve as a strong ultimatum to the international community. The latter, then, would have to make up its mind and promulgate an international regulatory system. Bergsten draws an analogy between international ocean regulation and the international monetary system. The latter, after its breakdown, was not restored until individual countries took the initiative and let their currencies float.

Bergsten's solution, based on an initial thrust at increased nationalism, may lead to increasing perturbation in the already disturbed and delicate international order of the oceans. A series of unilateral action on all sorts of levels may follow. However, let us assume, as Bergsten does, that an international agreement is reached, owing to the unilateral action by the United States. Under these conditions, Bergsten advocates the redistribution of social savings that would be inherent in the increased output. Since Bergsten believes that there are only contrived shortages, given the break in production control by a national effort, international regulation of ocean production would serve to block any such contrivance by individual firms or primary countries. The end result would be increased and more efficient output of ocean resources. These papers point up two issues that surround the analysis of the use of ocean resources. The first involves reliance on markets and the traditional neo-

---

[11] For an alternative view of the number of potential entrants, see Giulio Pontecorvo, "Reflections on the Economics of the Common Heritage of Mankind," *Journal of Ocean Development and International Law* 2(1974):203-216.

classical welfare assumptions to allocate the resources efficiently and to distribute income equitably. The second involves priorities—just how important are fish and deep sea minerals to the United States? The historical relationship between market economics and ocean resource use has been briefly explored elsewhere.[12] Similarly, the explosion in legislation dealing with ocean management in the United States in the 1970s has been catalogued, and its importance as the basis for government intervention in the markets for ocean resources has been noted.[13] The set of laws passed in the 1970s is important because it clearly extends government regulation of the use of ocean resources. Intervention in markets involves a choice between accepting a market-based solution to the efficiency-equity problems or accepting the premise that markets are imperfect and that the results they achieve are unacceptable. Here, we are not directly involved in the question of the desirability of planning or free markets, since the United States' commitment to planning, to some degree, is clear. What is important is that the observed price vector for ocean resources is not optimal. In these circumstances, the welfare conclusions provided by partial studies must be considered with great care.

The Nathan Report

An especially interesting study was undertaken by R. R. Nathan Associates, Inc., under contract to the Congressional Research Service of the Library of Congress.[14] The Congressional Research Service had acted in response to a request made by the National Ocean Policy Study of the Senate Commerce Committee for an appraisal of the recent and prospective economic value of ocean resources to the United States. The so-called Nathan Report attempts to measure the economic value—the gross value of output—from ocean resources. Emphasis is on the primary output of goods and services—essentially those performed in the ocean. The primary value is based on total revenues. That is, it represents the product of estimated unit prices and quantity of output at the first transaction point (e.g., minerals or fish delivered to a port). The concept of primary value excludes closely related processing, distribution, or other

[12] Pontecorvo and Wilkinson, "From Cornucopia to Scarcity: The Current Status of Ocean Resource Use."

[13] Pontecorvo and Wilkinson, "The Contribution of the Ocean Sector to the United States Economy." The most important of the federal laws are Fisheries Conservation and Management Act of 1976 and the Coastal Zone Management Act of 1972.

[14] U. S. Congress, *The Economic Value of Ocean Resources to the United States.*

functions performed at later points. However, such secondary values are estimated in the study where the available data permit.

In addition to, or apart from, the gross value approach, the concept of rent as a measure of economic value is developed. Unfortunately, this approach is considered outside of a more general theoretical framework of a value added approach to a system of national income accounting. The Nathan study denotes the rent as the extra profit derived from the nature of the resource and not from any special attributes of the party engaged in the related activity.

The Nathan study also utilized the concepts of opportunity cost and consumer surplus. Opportunity cost is defined as the value of a resource in terms of substitutes that cost more. Consumer's surplus is the difference between the price of a good or service and what the consumer would be willing to pay. If the latter is greater than the sale price, the consumer incurs a gain called the consumer surplus. For illustrative purposes, suppose that it costs more to extract a certain mineral from terrestrial sources than from ocean deposits. An industry that uses this mineral as a major input can sell its final product at a lower price if the ocean source is marketed at a lower price than output from a terrestrial source. Consumers demanding a certain volume of that product gain by satisfying the same amount of demand at a lower price. Note that the application of this concept assumes that perfectly competitive market conditions are satisfied.

Estimating consumer surplus is one basis from which to launch social policies aimed at encouraging the development of particular resources or activities. The concept of opportunity cost can be employed similarly. If the ocean resource costs more to develop than does its terrestrial counterpart, the industry will not have the incentive to engage in developing the ocean resource. The opportunity cost estimates, then, provide a yardstick for measuring the profitability of an activity.

My comments concerning the inadequate theoretical framework and accounting system within which to examine the concept of economic rent also apply to the use of the consumer surplus and opportunity cost concepts. It is unclear under what assumptions such concepts would be valid measures of the economic value of ocean resources. It is also unclear how they are used in the Nathan Report.

Despite its shortcomings, the Nathan Report is the most comprehensive and best organized among many empirical works on the value of ocean resources. It provides useful guidelines for assessing the dollar value of ocean resources and activites. The findings are, however, merely estimates. Given the level of reliability of the pertinent data, imprecise estimates and projections are to be expected.

The more serious question is the lack of a consistent methodology. The main problem in this context arises in inputting economic rent. The latter is often confused with opportunity cost and/or consumer's surplus. These three concepts are not identical, and none is clearly utilized in the study. Such estimates and projections must be revised before policy decisions can be formulated. The major value of the Nathan Report is in focusing attention on an area of growing national concern—the value of ocean resources and economic activity.

An Ocean Account

Is it possible to resolve any substantive aspects of the methodological problems posed in the literature? As noted earlier, the purpose of creating an ocean account within the framework of the existing National Income Accounting System is to deal with two issues: (1) to provide a consistent measure of the aggregate value of output from the oceans, and (2) to create a data base that permits the measurement of relative values in the several subcomponents of ocean activity within the theoretical and empirical limitations of the National Income Accounting System. This has been accomplished by a joint project of the Columbia University Business School and the United States Department of Commerce.

Substantive analytical issues are involved in the development of the methodology required to separate an ocean sector from existing national income accounts. These issues include: whether or not the method is general, that is, can it be used in some form to create other sectors, such as an agriculture sector; and enumeration of detailed steps, rules, and guidelines required to transform the existing national income data into a two sector (ocean/all other) account. The details have been presented elsewhere;[15] here we present only the broad outline of the method.

The existing accounting system measures the value added by each step in the economic process. Thus, Table 1 (which is given here on a relative basis) measures the value added in the system by industry. In the national income accounting system, underlying data are built up from the concept of an establishment. Similar establishments are grouped to form gross product originating (GPO) sectors, and ultimately these GPO sectors are aggregated to the industry level. The actual accounting data rest on census material, census of manufacturing and retailing, internal revenue data on earnings and estimates, and measures of the returns to capital and land. To create an ocean sector, it was necessary to reach

[15] *Pontecorvo, Wilkinson, et al.,* "Contribution of the Ocean Sector to the United States Economy."

down to the establishment level to see what portion, if any, of an establishment's output should be allocated to the ocean sector and what portion should be in the all-other category. This was done utilizing criteria developed for both the demand and supply sides of the establishment's operations. There were difficult problems in the decomposition of the government sector, since the level of detail of reporting is well below the information base in the private sector. At first approximation, this process of creating the ocean account may be visualized as, say, cutting a sausage vertically rather than horizontally, but where the size of the sausage must stay the same, that is, the value of the Gross National Product (GNP) must be invariant to any subdivisions created within it. The analogy is incomplete, however, since any restructuring of GNP requires a complete set of guidelines, rules, and definitional criteria that yield consistency. The elements that are combined to create the new sector must be selected in a manner that is consistent across the full range of economic activity.[16]

The process then involves the following steps; definition of the sector, establishment of consistent criteria for the selection of that portion of the output of an establishment that belongs in the defined ocean sector. Aggregation of these results yields the ocean sector components and, ultimately, the estimate of the ocean sector itself.

In the aggregate, the ocean sector is slightly less than three percent of GNP for 1972. At first, this may seem to be a modest portion of gross national product; however, given the size and complexity of the United States economy, it is comparable to other similar sectors. The ocean sector is approximately the same size as agriculture (3 percent), mining (2 percent including oil and gas extraction; note that price increases since 1972 have raised the relative importance of this sector), construction (5 percent), and transportation (4 percent). It is important to realize that the large portions of value added in the United States economy come from the transformation of output in manufacturing, in trade (wholesale and retail), and in the output of services and government.

While the value of output (3 percent), for the ocean sector is comparable to other extractive industries in the system, it is clear that we need to move beyond this figure and look at employment and the nature of the capital stock utilized by the ocean sector to get a more complete picture of its economic importance. Fortunately, the definition of the ocean sector required to measure the flow of income within the framework of the national accounts now provides the guidelines necessary for the next step, which is an important task for the future.

---

[16] There are, of course, numerous identifiable problems involving arbitrary definitions and both the availability and the adequacy of the data base itself. In most instances, these are problems affecting the income accounting system itself.

Table 1*

Value to the United States Economy of Output from the Ocean
1972 Data
Two-Sector Systems (Ocean/All-Other) from Gross National Product
by Major Industries

By Sector	U.S. Total: Industry Proportion of GNP Total (in percent)	Ocean Sector Total: Industry Proportion of Ocean Sector Total (in percent)
Ocean sector	3	100
Non-ocean sector	97	0
Total by sector	100	100

By Industry	U.S. Total: Industry Proportion of GNP Total (in percent)	Ocean Sector Total: Industry Proportion of Ocean Sector Total (in percent)
Agriculture, forestry and fisheries	3.0	1.0
Farms	2.7	0
Agricultural services, forestry, and fisheries	0.3	1.0
Mining	1.6	7
Oil and gas extraction	1.1	7
Mining and quarrying of non-metallic minerals	.1	0
Metals and coal	.4	0
Construction	5	1
Manufacturing	24.7	4
Non-durable goods	10	—
Durable goods	14.7	—

Industry						
Transportation	3.9			8		
Water		.2			8	
Other than water		3.6			0	
Services		0.1			0	
Communication	2.5			0		
Electric, gas and sanitary services	2.4			0		
Wholesale and retail trade	17.2			24		
Retail trade		9.9			24	
Wholesale trade		7.3			0	
Finance, insurance and real estate	14.3			15		
Real estate		10.5			—	
Finance and insurance		3.8			—	
Services	11.5			3		
Hotel, motel and lodgings		0.6			2	
Amusements and recreation		0.7			1	
Education		0.6			—	
Other		19.6			0	
Government and government enterprises	13.2			37		
General government		11.7			36	
Federal			4.3			35
State and local			7.4			1
Enterprises		1.5			1	
Federal			0.7			—
State and local			0.8			—
Rest of world	0.6			—		
Statistical discrepancy	0.1			—		
Total, by industry	100.0			100.0		

*All data are preliminary and are subject to possible revision. Source: Project, "The Aggregate Income and Product of the Oceans," prepared by the U. S. Department of Commerce and Columbia University. For details on the definition of the ocean sector and statistical methodology, see Pontecorvo, Wilkinson, et al., "Contributions of the Ocean Sector to the United States Economy."

Below the level of the national aggregates, we see the relative position of the various elements that go to make up the ocean sector. Again, it is clear that trade and government (primarily the navy) are the largest contributors. This reflects the crucial importance of the coastal zone, with its recreational facilities, within the ocean sector.

Concluding Comments

If we move from what is explicit in the table to the range of issues that are implicit, we open for ourselves a whole set of research questions. These issues are not new in an intellectual sense, but the availability of data now permits different **approaches** to policy and theoretical issues. For example, given the value of the output of fisheries, does it appear that the cost of fisheries management by federal and state governments is low or high? No simple answers are available for such questions, but the creation of the ocean sector account provides additional insight and a framework for analysis.

CHAPTER 10
LOCAL GOVERNMENT DIVERSITY AND FEDERAL GRANT PROGRAMS: DESIGN AND MODIFICATION OF THE COASTAL ENERGY IMPACT PROGRAM

ROBERT L. BISH*

Introduction

The design and implementation of intergovernmental programs to achieve national objectives is a difficult task. While the technical analysis of problems and conceptual design of solutions is often easily accomplished, the passage of solutions into law, the administrative implementation of laws in a politicized system, and the interface with potentially more than 80,000 state and local governments provides many opportunities for special interests to alter programs. Sometimes the result is distortion sufficient to prevent achievement of the original objectives; a more likely result, however, is simply diversion of some funds so that the program appears successful even if it is inefficient. This latter outcome may even be in the interests of the originally intended beneficiaries if the special interests to whom funds are diverted are politically strong or are strategically located in the legislative or administrative process so that they can make the entire program larger than it would have originally been.

The Coastal Energy Impact Program provides an excellent case of a problem analysis, program design, legislation, implementation, and modification of an intergovernmental program. Its lessons provide useful insight into the difficulties of undertaking and implementing technically designed programs in the American system.

* Robert L. Bish is Professor in the School of Public Administration at the University of Victoria, Canada. An earlier version of this paper was prepared for a conference on Formulating Marine Policy: Limitations to Rational Decision-Making, sponsored by the Center for Ocean Management Studies, University of Rhode Island, in cooperation with the Office of Ocean Management and the Office of Sea Grant of the National Oceanic and Atmospheric Administration, Narragansett Bay Campus, University of Rhode Island, Narragansett, Rhode Island, June 19-21. 1978.

The Coastal Energy Impact Problem

The increased prices of foreign petroleum products and the continued increase in domestic energy consumption have provided incentives for new and expanded production of off-shore oil and gas and the import of Alaskan crude and foreign liquified natural gas. The new energy facilities associated with each of these energy sources may have major effects on land use and the social, economic, and natural environment in the coastal zone.

If all energy facilities generated net benefits for the state and local governments within which they were located, there would be no reason for the national government to encourage the development of energy facilities. However, because oil production may take place off-shore and outside a state's taxing jurisidiction, or in a state or local government different from the one bearing the costs of providing public services to the associated population, it is possible for a state or local government to suffer a negative fiscal impact from an energy development in addition to whatever disruptions of the natural and physical environments are brought about by the facility. Recognition of the potential for adverse impacts that would lead state and local governments to refuse the development of energy facilities in their areas has resulted in legislation designed to overcome this disincentive problem.

This analysis will briefly describe the nature of fiscal impacts from energy facilities and analyze the program designed to deal with adverse impacts—the Coastal Energy Impact Program (CEIP). The focus of the analysis will be on how the diversity of state and local governments and their revenue mechanisms make a program that is simple in concept extremely difficult to translate into law and regulations. The resulting complexity of the law and regulations then provides an opportunity for particular interests to engineer apparently minor modifications in the law to obtain significant benefits beyond the original conception of the CEIP program. This pattern of program design followed by program distortion appears to underlie many criticisms of the growth of government. It may not be something that can be eliminated, but it is a factor that should be taken into account in evaluating the potential for efficient intergovernmental programs.

Fiscal Impacts

State and local governments obtain their revenues from taxes and user charges from individuals and businesses, or from grants and contracts from other units of government. The problem that arises with offshore energy development is that the business tax portion of a state or local government's revenue is not accessible because the development of the facility is beyond its taxing jurisdiction. Hence, it is possible that the revenues derived only from taxing individuals, without the business taxes, will be insufficient to cover the costs of providing public services to the people associated with the energy facility. While it was the issue of federal offshore oil and gas production that generated interest in the problem of negative fiscal impacts on state or local governments, it must be noted that any time an investment occurs in one jurisdiction but the associated population resides in another, there is a potential for a negative fiscal impact on the jurisdiction with the people and a tax windfall in the jurisdiction with the investment.

To support the idea that states bore negative fiscal impacts, both Texas and Louisiana sponsored studies to demonstrate negative effects.[1] These studies are cited many times in Congressional hearings, even though subsequent analysis indicated that each study was inaccurate and that positive fiscal impacts were more likely than negative ones for each state as a whole.[2] Further analysis, sponsored by the Office of Technology Assessment (OTA) for Congress, indicated that for a state and its local governments aggregated together, the general pattern for offshore petroleum developments was three years of net fiscal costs during early exploration stages, several years of high fiscal surpluses during construction of facilities, and then moderate fiscal surpluses into the future.[3] The initial costs were generated by the need to service populations during the time when virtually no onshore business tax revenues could be expected. The following large net fiscal surpluses were the result of sales and use taxes on construction inputs, real estate transfer taxes, and property taxes generated by the onshore components of the offshore activity. The continuing surpluses were the result of property taxation on the onshore components such as pipelines, tank farms, and the like. It should be noted that these conclusions were reached without inclusion of corporate income taxes on in-state activity (all coastal states but Texas and Washington have state corporate income

[1] *Benefits and Costs to State and Local Governments in Texas Resulting from Off-Shore Petroleum Leases on Federal Lands*, Office of the Governor, Austin, Texas, November 14, 1974; and *Off-Shore Revenue Sharing: An Analysis of Off-Shore Operations on Coastal States*, Gulf South Research Institute, Baton Rouge, Louisiana.

[2] Appendix A, "Other Fiscal Impact Studies," in Robert L. Bish, "Fiscal Effects of OCS Oil and Gas Development and Deepwater Port Development," *Coastal Effects of Off-Shore Energy Systems, Vol. II: Working Papers*, United States Congress, Office of Technology Assessment, November 1976.

[3] Bish, "Fiscal Effects of OCS Oil and Gas Development and Deepwater Port Development."

taxes) or business real personal property taxes on oil inventories. From the results of the OTA-sponsored analysis, it would appear that the major problem was not likely to be long-run net negative fiscal impacts because states and local governments have sufficient taxing mechanisms to capture revenues from on-shore components of the activity. However, there are still fiscal problems, including: (1) the timing of costs and benefits: costs must be incurred to provide services for people before any revenues from business taxation could be anticipated; thus, front-end costs could be high; (2) risk: a state or local government could borrow and invest in facilities to service people, and the oil discovery could be smaller than anticipated, leaving the government with bonds to pay but insufficient revenues; and (3) spatial mismatch: even though offshore developments may result in long-run positive revenues for all state and local governments, if the onshore service facilities are located in a state, it is still possible for it to bear net costs if the oil is landed in an adjacent state. An even more likely possibility is that a particular local government would end up bearing the costs of servicing people while the oil is landed outside its jurisdiction. This would be most likely to happen in areas where people are clustered in coastal towns but where the oil is landed or construction facilities are developed outside the town's boundaries. This problem is further complicated by the fact that most people reside within the jurisdiction of many local governments simultaneously, each of which may have different boundaries. Among the most common local governments are not only counties and cities, but townships, school districts, water districts, hospital districts, fire districts, postal districts, and parks and recreation districts. In addition, different local governments use many different taxing mechanisms such as the real property tax, personal property taxes, sales taxes, income taxes, payroll taxes, business taxes, and license fees and user charges. The likelihood of all of the people serving an energy facility and all of the onshore investments from the facility occurring in the same set of local government jurisdictions is very small in most parts of the country.

Recognition of the potential for negative fiscal impacts on some governments but not on others led to a clear conceptual framework for an energy impact program. Such a program would need a mechanism for lending to cover front-end costs which could be repaid. It must be able to absorb the risk for government investments when expected revenues do not materialize. It must compensate state and local governments where a spatial mismatch left them in a negative fiscal position. In addition, planning funds and compensation for damages to the natural environment and any reduction in recreational opportunities caused by energy facilities were viewed as further ways to reduce disincentives for energy facility developments. Recognition that in the absence of a spatial mismatch or risk, large positive fiscal benefits were likely, and a desire to avoid excessive coastal developments for the sake of capturing federal windfalls also led to concern for designing a program which would remove disincentives but not provide additional federal windfalls to state and local governments. This is a rather simple set of objectives that is well-matched to the nature of the problem.

Alternative Funding Mechanisms

Several funding approaches were considered in hearings and analyses.

Shared Revenues. Some congressmen, notably from Louisiana, strongly advocated a share of federal royalty revenues being returned to the adjacent state.[4] This approach was modeled after the federal return of a portion of mineral lease royalties from mining on federal lands. The shared revenues approach has several deficiencies. First, it does not deal with the problem of front-end costs. The costs of serving people would be incurred long before production-generated royalty payments. Second, there is no clear way to see that royalty revenues go to the state or to local governments which bear the costs of servicing the energy facility population. Royalty sharing could just be a windfall benefit, on top of already high projected positive fiscal impacts for most states. And third, it would be unfair to non-coastal states to share royalties with only coastal states in a way not related to costs because offshore lands are truly federally owned and governed and are not like federal mineral lands which are all within the boundaries of particular states.

Formula Grants. Formula grants could be based on a variety of factors, including factors which would provide front-end funds not possible with royalty sharing. The major problem with reliance on formula grants exclusively, however, is that it would be impossible to develop a formula

---

[4] U. S. House of Representatives, Subcommittee on Oceanography of the Committee on Merchant Marine and Fisheries, *Hearings on Bills to Amend the Coastal Zone Management Act of 1972*, U. S. Government Printing Office, 1975, p. 54. Louisiana has a special reason for favoring shared revenues or formula grants over project grants to remove disincentives to new energy facilities. Louisiana already has a developed structure for oil production both onshore and offshore. As onshore production declines, onshore workers and facilities will simply shift to offshore production, and few new facilities will be required. A sharing of royalties instead of project assistance would make more fiscal difference to Louisiana than to any other state.

that would provide revenues to negatively impacted state and local governments without simultaneously providing windfalls to others. The most difficult part of any formula would have been how to treat the diverse nature of local governments within each state, or how to evaluate any state-designed formula for pass-through to see if local governments bearing costs were actually compensated while others, including the state government itself, did not receive large windfalls. As will be seen, a component of a formula grant was provided in the Coastal Energy Impact Program for planning funds and environmental and recreational impact compensation.

Project Assistance. The most precise targeting of aid can be through individual project assistance. One can analyze any single energy facility impact and decide whether the affected government should receive a loan or a grant, determine the repayment schedule, and so on. This is the approach that could solve the energy impact problem without generating large windfalls. Under an ideal set of regulations, any local government official contemplating approval of an energy facility would know his legal grounds for receiving federal funds to offset any adverse impacts. Thus, the more specific the rules the safer he feels in anticipating federal decisions. At the same time, Congress, in drafting laws, especially laws which may dispense over a billion dollars, also likes to bind administrators into a set of rules so that administrative decisions are those desired by Congress. In short, a reduction in administrative discretion usually increases the predictability of the program. The difficulty with project assistance to one of 80,000 local governments, however, is that to reduce administrator discretion, the set of rules, including those which guide the forecasts of population change, public service costs and local government revenue charges must be extremely complex. Moreover, rules with sufficient scope to cover any possible energy facility impact in any possible local government jurisdiction would, on the face of it, be too complicated for any single local government official to understand, because only ten or twenty percent of the rules may actually apply to his situation. Both the resulting Coastal Energy Impact Program Law and the regulations attempt to balance the achievement of CEIP objectives with reduction in administrator discretion in a diverse environment. However, early congressional reports appear to have severely underestimated the difficulty of achieving this balance.[5]

[5] Jeffrey Roughgarden and Gerald Sauer, "Comparisons of Alternative Methods for Distributing Coastal Energy Impact Funds," Appendix 5 in U. S. Senate Committee on Commerce and the National Ocean Policy Study, *Energy Facility Siting in Coastal Areas*, U. S. Government Printing Office, 1975, pp. 121-126.

The Coastal Energy Impact Program Law[6]

The 1976 CEIP Law established two funds: one is for formula grants and is authorized $50 million a year for eight years; the other is the CEIP fund of $800 million for ten years, basically for project-type assistance. These two sources of financial aid represent a compromise between House of Representatives (Louisiana) advocates of revenue sharing and Senate and technical analysis recommendations for project-type assistance directly related to costs imposed on state or local governments by the development of new energy facilities. The formula grant fund authorized disbursement of money to states based on a formula taking into account the preceding year's federally leased acreage off of that state's shores, oil and gas produced, and oil and gas landed in the state from offshore lands, and the new employment attracted to the state by offshore activity. The primary use of formula grants is for planning and to alleviate damage caused to environmental or recreational areas. Revenues allocated via the formula grants could also be used as a secondary source of funds if CEIP fund allocations for project assistance are exhausted.

The primary use of the CEIP fund of $800 million is for project-type assistance. There is a formula allocation based on the costs of providing public services and anticipated population increases due to energy facility activity, but the allocation is simply the maximum a state may apply for on a project-by-project basis. The major stated intention for the CEIP fund is to provide loans and loan guarantees for the financing of public investments, and occasionally public services, prior to accrual of associated revenues from energy facility development. The provision of loans and loan guarantees takes care of the front-end financing problem associated with energy facility development.

The risk element of public facility development is also explicitly provided for by making forgiveness of any loan, or payoffs associated with loan guarantees, automatic if anticipated revenues did not materialize. The law stipulates that each project for which an application for assistance is made must include forecasts of population change, the costs of public facilities, and the revenues anticipated. It is clear that whenever anticipated revenues are sufficient to repay the loan that either a loan or loan guarantee is virtually automatic under the CEIP program.

Less clear in the law itself is the treatment of the spatial mismatch problem. It is implied that local governments which do not anticipate receipts to exceed costs should be aided, but it is also clear that funds for public facilities must be provided through loans or loan guarantees rather than grants. Only if revenues are then insufficient to retire the loan do

[6] *Coastal Zone Management Act Amendments of 1976,* P. L. 94-370, Section 308.

grants for repayment assistance become available. A question that is somewhat vague in the law is the administrator's discretion to grant loans when revenues to retire the loan cannot be forecast. It should be noted, however, the project-related assistance is flexible enough to deal with any kind of local government unit—a flexibility that is necessary because of the large number of local governments with their diverse functions in different areas.

The law is not a simple one, but it is directly aimed at achieving specific objectives and is probably about as simple a law as could be anticipated to achieve the desired results. If there is any source of confusion it is in the creation of two separate funds to accomplish four different purposes, with one fund allocated on a formula grant basis for a restricted list of activities, and the other mainly for specific project-related assistance up to a maximum amount for each state—also calculated by formula.

The other potential weakness in the law is the treatment of local governments. The law simply states that

> Each coastal state shall, to the maximum extent practical, provide that financial assistance provided under this section be apportioned, allocated, and granted to units of local government within such state on a basis which is proportional to the extent to which such units need such assistance. [Section 308 (g) (2)]

Furthermore, the degree of federal supervision of whatever process is developed is limited to review of following the process—not the results. Given the diversity in local governments, such an approach may be all that is possible, but it is not one that many local government officials felt comfortable with, and some of their discomfort was validated in early state plans. In Maryland, for example, the State Energy and Coastal Zone Administration included only counties and municipalities as eligible local governments, excluding special districts and special-purpose agencies except as they apply through a county or municipality.[7] This exclusion from direct participation is directly contrary to both the intent and specific wording of the CEIP Law and Regulations, and it will pose an interesting case for federal adjudication if it is challenged.

---

[7] Energy and Coastal Zone Administration, "Draft Interstate Allocation Process for the Coastal Energy Impact Program in Maryland," Maryland Department of Natural Resources, May 18, 1978, p. 2.

Coastal Energy Impact Program Regulations

Any law must be operationalized in regulations and administrative procedures. The development of regulations for the CEIP has been controversial. The controversy stems from several dilemmas. First, the strategy adopted for drafting regulations was to be sure that restrictions, statements of administrator discretion, and controversial points were placed in the first drafts so that state and local officials would have an opportunity to comment prior to publication of final regulations.[8] This approach was considered superior to that of publishing brief and vague initial regulations for comment and then putting all the restrictions in the final regulations—when it is too late for comments and revisions. While the approach taken may have produced the desired results, it caused considerable controversy for program administrators and provided opportunities for advocates of pure shared revenues who disliked the attempts to limit windfalls to try to shape the program to fit their preferences or force a rewriting of the law itself. Most of these attempts were quite blatant and did not have a major impact on the initial final regulations issued following the hearings.[9] They did, however, have an impact on the final regulations, especially in reducing restrictions on the uses of formula grant funds and the elimination of restrictions to *new* energy facilities of loans, guarantees, and repayment assistance from the CEIP fund.[10] Both of these changes clearly responded to concerns of Louisiana congressmen, although the elimination of the restrictions of

[8] Department of Commerce, NOAA, "Coastal Energy Impact Program: Proposed Regulations for Financial Assistance to Coastal States," *Federal Register Part II*, Friday, October 22, 1976. *Idem.*, "Coastal Energy Impact Program, Interim-Final Regulations for Financial Assistance to Coastal States," *Federal Register Part III*, Wednesday, January 5, 1977; and *Idem.*, "Coastal Energy Impact Program: Administrative Procedures," *Federal Register Part IV*, Thursday, February 23, 1978. There were also at least three unpublished drafts of regulations that were circulated to Congress and to state and local public officials, but the three published versions permit analysis of changes and responses to comments from one draft to another.

[9] Hearings before the Subcommittee on Oceanography of the Committee on Merchant Marine and Fisheries, House of Representatives, 94-2, *Oversight on Regulations Being Proposed by the Office of Coastal Zone Management Act Amendments*, December 10, 1976. These hearings represent a rather strong "congressional" attack on the "restrictiveness" of the regulations. "Congressional" must be interpreted carefully, however, since only six members of the 23-member House Subcommittee on Oceanography attended and only three played an active role: Chairman Breaux (Louisiana), Treen (Louisiana), and DuPont (Delaware). Each of these three was an advocate of pure revenue sharing, and Breaux and Treen were surely aware that only with revenue sharing without a focus on *new* energy developments, would Louisiana be a major beneficiary of the CEIP program.

[10] Department of Commerce, National Oceanic and Atmospheric Administration, *Federal Register Part IV*, Thursday, February 23, 1978, pp. 7546-7.

loans, guarantees, and repayment assistance to projects necessitated by *new* or *expanded* energy facilities is clearly contrary to the intent of the legislation and the rationale for the CEIP program to stimulate energy production.[11]

A second dilemma was the simple tradeoff between administrator discretion and certainty for state and local government officials, complicated by the difficulty of actually forecasting the fiscal impacts of energy facilities on a local government. In some areas, administrator discretion was reduced in subsequent drafts of the regulations. For example, the automatic nature of repayment assistance when revenues to repay loans did materialize was clarified. However, in other areas administrator discretion had to be increased because rules developed to account for each potential local government situation would have been so complex that no one could have understood them. The increase in administrator discretion was in the most critical project evaluation area— that of deciding on the "quality" of forecasts of fiscal impacts, the repayment terms for loans, and what conditions should accompany loans when revenues for repayment cannot be forecast. This is likely to be the most critical area, because if there is no spatial mismatch almost any forecasting technique will predict adequate revenues to repay loans, and sufficient revenues will be collected for this purpose. However, when repayment is so assured, it is better for the local government to borrow through the municipal bond market or use a state-sponsored borrowing program because the interest rates on the loan will be lower than that of CEIP loans or under CEIP loan guarantees, because loan guarantees eliminate the tax-exempt status of municipal bonds and, hence, higher interest rates will have to be paid. It is precisely the marginal cases where forecasts are uncertain or where revenues are unlikely to be generated that a local government would have to use CEIP borrowing if it were to obtain funds at all. And it is precisely here where administrator discretion remains paramount.

A serious attempt was made to reduce administrator discretion in this area in the preliminary regulations by specifying the components of a simple fiscal forecasting model developed specifically for CEIP purposes.[12] The model would have required historical data on revenues, expenditures, and population changes for the previous ten years, and

[11] For an analysis of the "new and expanded" requirement, see *Oversight Hearings,* pp. 26-28, and Interim-Final Regulations of January 5, 1977, p. 1167.

[12] Robert L. Bish, John D. Wolken, Candis L. Brown, and OCZM staff, *The CEIP Impact Model: Technical Assistance Materials;* Robert L. Bish, John D. Wolken, *The CEIP Impact Model Technical Manual;* and Robert L. Bish and Candis L. Brown, *Issues in Energy Facility Impact Forecasting.* All prepared for the Office of Coastal Zone Management, NOAA, Contract No. 7-35174, 1977.

would then utilize existing tax rates to forecast population, revenues, and expenditures with the energy facility impact. A maximum amount of the difference between revenues and expenditures would have been specified in the regulations for the payback schedule, and if revenues were insufficient a payback schedule could be designed in anticipation of receipt of repayment assistance when it was verified that revenues would, in fact, be insufficient to cover costs. The entire purpose of this process was to reduce the discretion of the administrator and provide local government with a simple tool for forecasting energy facility impacts. Instead, several local government officials responded to the proposed regulations by indicating that such forecasts were too costly, and that too much data was to be required. Hence, specific forecasting requirements were eliminated from the regulations, and forecasts and payback schedules are negotiated on a case-by-case basis. The conditions for forgiveness of loans, however, were to be specified in the original loan agreement. Thus, a major effort to reduce administrator discretion and provide a certain position from which local government officials could begin their negotiations was eliminated at the request of local officials, but the conditions for repayment assistance were clarified prior to accepting the loan.

There are reasons beyond the costs of implementation for not relying on specific energy facility impact forecasts for future decisions. Small area fiscal impact forecasting is simply not a well-tested art. In spite of hundreds of thousands of dollars spent by the National Science Foundation and the Bureau of Land Management on energy impact forecasting procedures, none of the procedures can be disaggregated down to the local community level.[13] Thus, the procedure developed for the CEIP was unique, but it would have been desirable to have tested the method prior to using it as a basis for writing financial contracts.

Another area where administrator discretion was expanded, in response to oversight hearing criticism, was in the granting of loans to local governments for which repayment cannot be forecast. The law seems to indicate that such governments should receive aid, but the focus on loans instead of grants for public facilities makes it unclear what the administrator's position should be when repayment cannot be anticipated. The program administrator has specifically testified that loans will be available to impacted local governments even if repayment cannot be forecast.[14] In these cases, a deferred payment plan will be negotiated with repayment assistance automatically available when necessary. Because of

[13] The Weston Corporation has completed a detailed assessment of energy impact forecasting procedures under contract to the National Science Foundation, and these reports are in the process of being published. Some of their conclusions are summarized in Bish and Brown, *Issues in Energy Facility Impact Forecasting.*

[14] *Oversight Hearings,* p. 31.

the potential for the spatial mismatch problem, this area of discretion remains one of the most important in the administration of the entire program.

Modifications

The CEIP began operation in 1977. The requests for project-specific loans, and loan guarantees from the CEIP fund have been lower than anticipated because interest rates are higher than those generally available through direct municipal or state borrowing, and some areas of potential oil development, such as the Baltimore trough, have not been as rapid as anticipated. Except for CEIP officials' requests to the Treasury Department for lower interest rates on loans, however, little demand for change in its operation has appeared. The same cannot be said of the formula grant aid.

Formula grant funds are clearly preferred by state and local governments because they do not have to be repaid. Regulations specifying the use of such funds were originally quite specific, but following the oversight hearings, restrictions were relaxed to permit expenditures of virtually any public service or facility relatable to serving people or businesses associated with energy production. The formula for distribution to states was one-third related to offshore acreage leased, one-sixth each for oil produced offshore and oil landed in the state, and one-third for new employment attracted to the state by offshore activity. Of the $10 million appropriated (only $10 million was appropriated of the $50 million authorized) Louisiana was allocated $5.5 million and Alaska, Texas, and California each were allotted approximately $1 million. Smaller amounts went to Mississippi, Florida, and Alabama. The 1978 allocations were similarly concentrated, with $14 million of the $17.7 million appropriated going to Louisiana and lesser amounts to Texas ($2.4 million), California ($0.5 million), Mississippi ($0.3 million), Alaska ($0.2 million), and Florida ($6,775).[15] While the CEIP program was begun under regulations that accommodated concerns of those congressmen active in the oversight hearings, there was still interest on their part in moving toward straight revenue sharing and eliminating in law, as well as in the regulations, the focus on new or expanded energy facilities. During 1977, hearings to revise the procedures for the leasing of offshore lands by the Department of Interior provided an opportunity also to amend the CEIP program. In the hearings themselves, however, the senate

[15] Data from "Coastal Energy Impact Program: Allotments," 1/6/78.

committee did not consider revision of the CEIP program.[16] In House committee hearings as well, most of the emphasis was on offshore leasing procedures, but Congressmen Breaux (Louisiana) and Treen (Louisiana) did raise questions about CEIP procedures.[17] These were the same representatives who were critical of CEIP regulations in the oversight hearings and were the most vocal exponents of shared revenues and elimination of the restrictions of aid to public expenditures due to new or expanded energy facilities only.

During 1977, time expired before the Senate and House could agree on a single bill; consideration resumed in 1978, and the process was completed. Senate concern focused on offshore leasing, with the exception of Senator Johnston's (Louisiana) attempts to obtain sharing of offshore oil revenues and to eliminate the new and expanded provision for CEIP aid. Johnston's activities in this regard were somewhat unusual, since the committee considering the offshore leasing revisions was the Committee on Energy and Natural Resources, which does not have jurisdiction over the Coastal Zone Management Act of which the CEIP is a part. While Johnston's proposals were ultimately accepted by members of the important Commerce Committee in the Senate, the proposals did not have committee scrutiny by the original drafters of the CEIP legislation as is normally part of the amendment process.[18]

The House committee, under Breaux's guidance, did devote considerable effort to examination of the CEIP program. While straight revenue sharing was rejected, an increase in the authorization for formula grants from $50 million to $200 million per year was recommended along with a proposed change in the formula to a 50-50 formula based on leased acreage and landed oil and gas. This change in the formula would reduce the emphasis on new activity in favor of current production. It would also provide large windfalls because it is precisely where oil and gas are landed that the tax revenues accrue. Finally, the committee recommended that all requirements associated with aid only for public expenditures caused by new or expanded energy facilities be eliminated.[19] These provisions

[16] U. S. Senate, Committee on Energy and Natural Resources, *Outer Continental Lands Act Amendments of 1977*, 95-I, U. S. Government Printing Office, 1977, and *Idem., Outer Continental Shelf Lands Act Amendments of 1977*, Report 95-284, U. S. Government Printing Office, 1977.

[17] U. S. House of Representatives, Ad Hoc Select Committee on Outer Continental Shelf, *Outer Continental Shelf Lands Act Amendments of 1977*, Part 2, U.S. Government Printing Office, 1977, and *Idem., Outer Continental Shelf Lands Act Amendments of 1977*, Report 95-590, U. S. Government Printing Office, 1977.

[18] *Congressional Record — Senate*, July 14, 1977, S11860-S11869, and August 22, 1978, S13997-S13998.

were passed by the House. These changes would have greatly increased both the relative share and total amount of Louisiana's grants.

The conference committee to reconcile House and Senate bills did not accept all of the House provisions on CEIP amendments. The amount authorized for formula grants was increased from $50 million to $137 million (instead of the House-approved $200 million)—but with a 37-percent restriction on the maximum amount to go to any single state. While the increased authorization could help, the 37-percent limit clearly hurt Louisiana. The conference committee formula revision kept the House elimination of new employees but reduced the role of landed oil and gas to one-fourth and added offshore production (one-fourth) to benefit states where offshore activity occurs but where oil is landed in an adjacent state. This change reduced some windfall effects of the House proposal. Finally, the conference committee did eliminate the requirements that CEIP aid be for public expenditures that were caused by new or expanded energy activity only, and the conference recommendations were passed.[20]

While the description of the politics of CEIP revision, including analysis of several attempts from the floor to amend the bill has been greatly compressed, the forces for modification of the program are clear. The CEIP fund project-specific aid law and regulations provided for few windfall benefits, although its combination of loans, loan guarantees, and grants did make aid available for dealing with the problems of timing of costs and revenues, risk, and spatial mismatch of costs and benefits for new energy facilities. This was the original intent of recommendations derived from the technical analysis of the coastal energy impacts problem.

The development of the formula-grants part of the CEIP program, however, illustrates a classic fight over the distribution of windfall benefits and the use of political processes to modify a program to provide windfall benefits to a very few recipients. Virtually all activity for expanding this component of the CEIP program was by a few congressmen whose states would benefit significantly, while other congressmen, whose states were basically unaffected, showed little interest.

[19] *Congressional Record — House*, January 25, 1978, H203; January 26, 1978, H311, H322, H327, H266-7, H283-4, H288, H303-4, H311, H322, H327. February 2, 1978, H580-H589.

[20] *Congressional Record — House*, August 17, 1978, H8874-H8879, and *Public Law* 95-372, September 18, 1978.

Conclusions

The development of the Coastal Energy Impact Program in response to the national objective of removing disincentives for energy facility location provides several lessons for the design of federal grant and other intergovernmental programs. First, even though the conceptual framework to reduce disincentives to energy facility location was very simple, the diversity of state and local government structure, function, and finance would make so complex a set of rules that would really reduce administrator discretion so complex that no local governmental official (or most congressmen) could understand them. Thus, a complex program which includes extensive administrator discretion is necessary to achieve program objectives. This dilemma is being increasingly recognized in other areas as well.[21] Second, while the CEIP law and regulations limit windfalls from the project-specific CEIP fund, windfalls from coastal energy facility development are still very likely. As the study sponsored by the Office of Technology Assessment indicates, large revenues are likely to accrue from the onshore components of offshore development. With the spatial mismatch problem, it is quite likely that some local jurisdictions will receive tremendous windfalls (Calvert County in Maryland, for example, has gone from one of the poorest counties in property tax base per capita to one of the richest due to the Calvert Cliffs nuclear generating plant) while adjacent jurisdictions are receiving CEIP assistance. This kind of mismatch, however, is the province of state government and state constitutions, and the federal government has no authority to force changes to resolve this issue. Another source of windfalls, however, is the formula grant part of the CEIP program itself. It is likely that the revenues from this program will accrue directly to the same governments which receive tax revenues from the onshore components of offshore energy activity. The emergence of formula grants as a larger source of aid than the CEIP fund also illustrates a third important lesson from CEIP development: politics are a critical part of the development of any program, and partisans of special interests will attempt to get what they can from the federal treasury. One should note, however, that the increased emphasis on formula grants did not eliminate the CEIP fund and project-specific aid. Thus it is likely that the CEIP program as a whole can, with careful administration achieve its original objectives of resolving the problem of front end financing of new public facilities needed to service populations associated with new energy

---

[21] Robert D. Reischauer, "Governmental Diversity: Bane of the Grants Strategy in the United States," in Wallace B. Oates, *Political Economy of Fiscal Federalism.* (Lexington, Mass.: Heath Publishers, 1977) pp. 121-122.

production, eliminate the risk to governments for undertaking such investments, and compensate for losses due to spatial mismatch of public sector costs and revenues. The net result from the CEIP as a whole is a potential for achievement of original program objectives, but at a relatively high cost. Such an outcome should be expected in a complex federal system.

CHAPTER 11
SEAPORTS AS PUBLIC ENTERPRISES: SOME POLICY IMPLICATIONS

WILLARD PRICE*

Introduction

This chapter is concerned with seaport management and the concept of public enterprise. It is based on research conducted by the author and supported by the federal Sea Grant institutional program.[1]

I contend that a meaningful perspective for the study of seaport policy and management is through the concept of public enterprise. A model of public enterprise and a critique of the independence of these enterprises from the authority of other governmental units will be developed. A general research framework is suggested as a basis for empirical examination across the spectrum of seaports as well as those other activities which are considered public enterprises.

Seaports will be reviewed from the executive or management perspective, and not necessarily from a political science orientation or a maritime policy focus. I am less concerned with federal policy for the maritime industry than with the policy issues relating to seaports as platforms for maritime, commercial, and public activities. I will emphasize the implications of public concepts for the seaport manager. This discussion can contribute to a new professional development focus for seaport managers by providing an improved understanding of their environmental context and the policy options available to seaports.

To begin to construct a theory for public enterprise, it is necessary to introduce a range of enterprises across a public/private continuum. The following description is adapted from a presentation in a research

* Willard Price was Associate Professor of Public Administration at the University of Southern California at the time this was written. He is now Associate Professor of Public Administration at University of the Pacific.

[1] The University of Southern California's Sea Grant Institutional Program provided funding during 1978-79 for the author's research on "Curriculum Development Research Study to Support an Academic Field in Harbor/Port Management." Special assistance was provided by Helen J. Muller, doctoral student in public administration, who researched the public enterprise literature.

PUBLIC-PRIVATE CONTINUUM

Government Agency/Department
Under executive, budget, and personnel control of the government

Government Public Enterprise
Government owned, operated with some degree of independence

Mixed Enterprise
Governmental interest through partial ownership of equity in corporation

Private Enterprise
Corporate ownership held by private individuals

conference on public enterprises conducted by the Federal Executive Institute.[2]

The distinction to be made for public enterprises is that they remain government-owned entities, although they may experience some amount of independence or autonomy from other government entities. A basic variable across the above continuum is ownership, shifting from collective ownership of governments to the individual(s) ownership of corporations. Government subsidy and regulation can be involved in all four models, without restructing this range of concepts. Since government ownership infers authoritarian control by the public through political representation, one critical question is whether we expect less public control over policy and management when we create identifiable public enterprises.

There are several definitions of public enterprise, all with different emphases on ownership, management independence, subsidy, monopoly, and public accountability.[3] The Federal Executive Institute Conference began by formulating this broad definition:

[2] Frank P. Sherwood, ed., "General Group: Summary Presentations and Discussion," *Proceedings of a Research Conference on Public Enterprises* (Charlottesville, Va.: Federal Executive Institute, June 1, 1978), p. 77.

[3] William C. Shepherd, Public Enterprises: Economic Analysis of Theory and Practice, chapter 3, "Objectives, Types and Accountability" (Lexington, Mass.: Lexington Books, 1976), pp. 33-47.

. . . important government activities are being performed by public organizations which lie outside the classic bureaucratic structure. They are sometimes considered para-state, third sector, or quasi-public . . . the organizations arrangements differ vastly, as do systems of accountability and provisions for management discretion . . . (providing) various types of utility services, transportation, and financing and economic development . . . (a common element) is the commitment to serve a public interest. They are typically not profit-oriented and yet it is not always clear how to establish public accountability for these types of organizations.[4]

Annmarie Hauck Walsh, who has produced one of the most comprehensive books on public enterprise, offers this description:

Government (enterprises) are created by a public statute that defines their powers. They are owned by the government that so established them, but they are hybrid creatures, possessing some of the characteristics of private firms and some of public agencies. They are corporations without stockholders, political jurisdictions without voters or taxpayers . . . (providing) a relatively independent base of operations for entrepreneurs in the public sector. . . .[5]

The issue of independence is clearly the most frequently debated issue in public enterprise; managers seek the advantages of independence while certain public interests seek to restrict that autonomy in order to influence the policies and programs of the enterprise. The more liberal view is to keep the "public" in front of the "enterprise." Accordingly, a definition is suggested which recognizes independence as the key aspect:

Public enterprise can be defined as a focused public (government owned) function with some degree of independence in governance, management and finance.

"Focused" implies that the enterprise will be established to accomplish a single function such as seaports/airports, mass transit, power, water resources, housing, or even hospitals and other human services. In general, public enterprises are not defined as multiple-purpose governments or multiple-purpose special districts.[6] This last definition is

---

[4] Sherwood, *Proceedings of a Research Conference*, "Introduction," p. 6.

[5] Annmarie Hauck Walsh, *The Public's Business: The Politics and Practices of Government Corporations* (Cambridge, Mass.: MIT Press, 1978), pp. 3-4.

[6] Examples of public enterprises with multiple objectives include the Tennessee Valley Authority (TVA) and the Port Authority of New York and New Jersey. Both would argue that they simply have one broad purpose. TVA's would be economic development and the port authority's would be transportation and commerce development. Does the public have sufficient influence in the choices available among competing development alternatives?

intended to apply to enterprises at all levels of government, although this discussion will be limited to the local level, where many seaports are operated.

Current Attention to Public Enterprise

The public has options regarding the delivery of particular services. If it wants to exercise direct control through government agencies, it can hire government employees to deliver those services. A second option is to retain control over the planning of the service while contracting for service delivery from the private sector (or from another area of government) where the extent of control depends on the power of the contract document to affect the service performance. Less control is exercised in a third option when a government chooses to let the private sector deliver the service, but under government regulation. When monopolies exist, such as public utilities, the government can retain significant regulatory influence over the planning of services and the prices of those services. Finally, if the government chooses not to exercise direct influence over service delivery, then it must rely on the marketplace. It is expected that the public still desires competitive and efficient services under any choice of delivery method.

The question of which option is preferable is not a simple one, since the issue is clouded by alternative values. A more conservative person would normally prefer the latter two options, thereby emphasizing the free enterprise concept. The more liberal viewpoint would favor the first two options. Since this theoretically permits more public control of the result, the debate over "social balance" is, again, a political issue with many arguing that the size of the public sector or the relative size of the public versus the private sector has risen to a point where it threatens our basic economic and democratic system.[7] The counter arguments that the public sector is necessary to provide accountability and control, to protect the public against monopolies, or to subsidize desirable services that are not self-supporting are no longer as acceptable as they were before the current economic trauma of inflation and rising levels of taxation.

The current challenge the "publicness" is taking is in the form of a

Also see Robert B. Hawkins, *Self-Government by District: Myth and Reality* (Stanford, Ca.: Hoover Institutional Press, 1976).

[7] Bernard Herber, *Modern Public Finance: The Study of Public Sector Economics*, chapter 1, "Resource Scarcity and Intersector Allocation" (Homewood, Ill.: Richard D. Irwin, 1971), in which social balance is discussed. A review of the implication of tax initiatives and reduced government revenues is found in Charles H. Levine, "More on Cutback Management: Hard Questions for Hard Times," *Public Administration Review* 39 (March/April 1979):179-83.

serious political resistance to general taxation.[8] It is not yet clear whether or not these forces want to (1) reduce the property tax burden, (2) reduce the size of the government sector, or (3) eliminate selected programs. One significant question which has not yet been answered is whether those public services, including public enterprises, currently supported by fee systems, are immune from the new public resistance to taxation. Tax-resistance crusaders argue that increased fees or new fees for services previously paid by general taxation are, in effect, creating new revenues in violation of the intent of tax limitations.[9]

As a result of the financial stress on governments, new attention is being directed to public enterprises because they can demonstrate finance independence from general taxation. If tax limitations result in government establishing public enterprises that are supported by fees, the political acceptability of new or increased fees will still need to be determined. Services such as seaports, which have traditionally been limited through fees and without general taxation, are not as vulnerable to revenue reduction when they are not causing new taxation for the general public. At the same time, if the public concern is efficiency in government, will it view public enterprises as being able to operate efficiently even though they remain publicly owned?

Whenever governments identify functions to be delivered under public ownership, it is common to hear negative criticism about their performance. The conventional wisdom is that government employees and managers are not motivated to high performance, and that political leadership interferes with the objective of efficiency. One political response to this presumed inefficiency is to seek contracts with the private sector or to establish public enterprises. In both cases, the objective is to separate the service from conflicting governmental influence and to allow the management to perform efficiently, in effect, to be businesslike.[10] For public enterprises, the objective of independence and managerial freedom is granted to a public activity that has been deliberately set free (to some

---

[8] John J. Kirlin and Jeffrey I. Chapman, "California State Finance and Proposition 13," *National Tax Journal*, supplement, 32(June 1979):269-275; Willard Price, "Public Works in the Era of Reduced Revenues: 78-79 Impacts on PW Budgets in Selected California Cities," *Proceedings, 31st Annual California Transportation and Public Works Conference* (Institute of Transportation Studies, University of California, Berkeley, April 1979), pp. 59-78; and Maureen Fitzgerald, "Jarvis II," *California Journal*, 11(January 1980):35-36.

[9] In California many local governments have increased fees in the face of the property tax limitations required by Proposition 13. Several lawsuits have been filed challenging fees as a violation of the intent of Proposition 13. Their outcomes may affect the ability of some public enterprises to raise fees. The new tax initiative in California, Proposition 9, which advocated cutting the state income tax in half, was defeated in the June 1980 election.

[10] Walsh, *The Public's Business*, p. 255. Also see Harold Seidman, *Politics, Position, and Power: The Dynamics of Federal Organizations* (New York: Oxford University Press, 1975).

extent) from government. In effect, the enterprise itself seeks a relationship with other governmental bodies that is equal rather than subordinate. The implied task for public enterprise management is to deliver the authorized function efficiently and to remain financially viable.

Most public enterprise managers accept this mandate; they expect the freedom to manage, and are willing to stand accountable for financial success. Not all public enterprises, of course, can be financially self-sufficient, since they are neither able nor willing to raise sufficient revenues. If such enterprises are to survive, other governments will need to provide subsidies. The public purpose of subsidy may involve a desire to (1) keep the service available at reasonable prices, (2) provide capital financing not otherwise available, (3) maintain control of policy objectives, or (4) stimulate economic development when only the government is willing to act.[11] In these cases, the public purpose must be identified and accepted, particularly if taxpayers are going to continue to support the use of general taxation.

Possibly new conservative politics at all levels of government will not continue support for subsidized enterprises, although the public's choices may be limited for those services which cannot survive without subsidy. If the private sector is not a reasonable alternative for delivering necessary public services such as a mass transit, water resources, and airports and seaports, then a public enterprise or a mixed enterprise may be the only option.[12] With competent management, available resources, and a predictable environment, most seaports can survive as public enterprises. The question is whether or not they satisfy the public interest as they achieve independence and financial success.

The reality of public enterprise, in general, is not simply financial success, and managers should not expect to work under a one-variable standard of accountability. The "publicness" will remain, even if public enterprise may experience a greater degree of independence than other public programs. There will always be competing interest groups who will disagree on the objectives and development projects of the enterprise. These groups can only be served by broad political access to decision-making.

[11] Sherwood, "Small Group Discussion: Transportation Enterprises," and "General Group: Summary Presentations and Discussion," pp. 35, 85, 86.

[12] In Brazil, for instance, some seaports are mixed enterprises, with the federal public enterprise (PORTOBRAS) owning a majority share in those private companies operating ports. Mixed enterprises may offer a new compromise in the United States between enterprise independence and public interests. A majority ownership can allow control of policies and plans with minimal influence on the management of the enterprise.

The Paradox of Public Enterprise—Public or Enterprise?

There is a paradox within public enterprise that blurs the popular vision of the organization as an efficient and independent corporate entity.[13] To achieve the expectation of efficiency, it is implied that such enterprises must be more enterprise than public. But if enterprise is expected to be dominant, then a more appropriate model would be mixed enterprise or regulated private enterprise.

In public enterprise, the public focus remains as the primary emphasis, even though there is an opportunity for increased attention to enterprise management. The order of these terms distinguishes public enterprise management from business administration and suggests an opportunity for the profession of public administration. If the field of public administration is to be concerned about the management of public enterprise, it must revive a theoretical and empirical research base from which to shape a new emphasis in the professional educational programs of public administration.[14] This emphasis will likely require special attention to the management processes, finance, analysis, and performance of public programs. Nonetheless, enterprise managers cannot expect to neglect the public values, politics, ethics, and accountability perspectives that are normally emphasized by public sector education. While business administration's traditional subjects of production systems analysis, accounting, finance, and marketing are likely to be more relevant to public enterprise, it does not follow that the publicness can be neglected as the enterprise is emphasized. The challenge belongs to public administration to find the appropriate balance of these apparently conflicting topics.[15]

[13] For an early discussion of this concept, see Herman C. Pritchett, "The Paradox of the Government Corporation," *Public Administration Review* 1(September 1941):381-89.

[14] Public administration in the United States developed a literature on public enterprise in the 1940s and 1950s. See Harold Seidman, "The Theory of the Autonomous Government Corporation: A Critical Appraisal," *Public Administration Review* 12(Spring 1952):89-96; and Seidman, "The Government Corporation: Organization and Control," *Public Administration Review* 14(Summer 1954):183-92. The subject was almost neglected in the 1960s and 1970s, but a new opportunity comes with the 1980s. The recent aforementioned work of Sherwood at the Federal Executive Institute and the Walsh and Shepherd books are valuable foundations for another research forum on public enterprise.

[15] Sherwood, "Small Group Discussion," p. 45, and "General Group Discussion," p. 94. Many public enterprises are recruiting managers from MBA programs because analytical skills and a business sense are provided in MBA degrees. While MPA programs are adapting curriculum to include analytical content, they might also provide a special balanced curriculum for public enterprise managers, combining public issues with quantitative and entrepreneurial concepts.

A Research Framework for Public Enterprise

It is suggested that research for public enterprise should concentrate on the question of independence. No attempt will be made here to discuss the comprehensive research necessary to compare independence across functional, political, international, or historical variables.[16] Rather, an attempt will be made to develop a generic model by which independence in any public enterprise can be examined, although the only application will be made for seaports. First, independence or autonomy (no distinction in meaning is intended) will be defined as: the ability to make decisions on any aspect of policy and management without approval by higher authority.

To determine more specific measures of independence, several variables are selected by which independence can be observed across three categories: *governance, management and finance.* Each of these three categories will be defined by selected organizational decisions they include.

Governance

1. Establish the form of governance, including whether higher authority is vested in another (parent) government
2. Determine objectives, policies, and general plan
3. Approve development projects and services to be offered

Management

1. Establish the personnel system and select, promote, and terminate employees
2. Select technology, conduct engineering designs, and approve methods of service delivery
3. Approve lease agreements and contracts for goods and services

Finance

1. Establish fees for services
2. Select the accounting system and prepare separate accounting reports
3. Approve debt financing and disposition of surplus funds

Autonomous governance can be conducive to independence in management and financial decisions. On the other hand, it is possible, even ideal, for public enterprise to have more independence in

---

[16] One comparative review is available in W. G. Friedmann and J. F. Gormer, eds., *Government Enterprise: A Comparative Study* (New York: Columbia University Press, 1970).

management and finance than in governance decisions. In seaports, independent governance by a business-oriented board may be politically unacceptable if the board does not allow a broad-based public control over policies and plans.[17]

Further, enterprise governance boards will need to decide whether managerial and financial decisions can be delegated to executives or whether some or all of these decisions must be approved by the governing board. All governing boards must choose priorities for their influence. They may choose a narrow control stance and provide strict review of personnel, engineering, contracts, leases, fees, and accounting. But if they want to limit their influence to the governance questions established in the model proposed above, then they should concentrate their overview on policy, planning, and development.

Whether the governance authority is held by an independent board or by another constituent assembly (i.e., city council), the *normative theory of public enterprise* suggested here would require seaport executives to retain the *maximum amount of managerial freedom and a moderate amount of financial freedom.* Maximum managerial freedom to decide on the port's delivery systems and personnel policies provides the independence that enterprise managers expect as they try to deliver services efficiently. Only moderate financial freedom is expected; the policy and development actions of governance decisions impact seriously on financial results, since public interests may conflict with the single objective of enterprises to maximize business and financial performance. This theory of independence will be discussed in more detail relative to seaports in a subsequent section.

Another important question is the relationship of independence to organization performance. The theory above will not be extended to imply that management independence will maximize performance. Further research is needed to determine the relationship of performance measures and observations of independence across governance, management, and finance. But this question must be addressed before students of seaport management and public enterprise can evaluate the utility of independence.

[17] A strong case for the public side of public enterprise is made by Walsh, *The Public's Business,* pp. 343-56; also, Henry Tulkens, "The Publicness of Public Enterprise," in Shepherd, *Public Enterprise,* pp. 23-32. Also, for a negative view of independence, see Ronald C. Moe, "Government Corporations and the Erosion of Accountability: The Case of the Proposed Energy Security Corporation," *Public Administration Review* 39 (November/December 1979):566-71.

Selected Policy Issues in Seaport Management

A seaport should be understood as a bridge between the sea and the land; in essence, it is a platform for the transfer of material from a water mode of transportation to a land mode. The interface of one mode of transportation with another is the basis of the generic meaning of "port." Seaports, like airports, are essentially the facility for this transportation interface. United States ports commonly act as housekeepers whose principal task is to facilitate commercial activities by developing and maintaining piers, cranes, terminals, warehouses, and other facilities.

In addition, seaports have choices of the types of activities which they allow within their jurisdiction. These range from commercial maritime activities to public areas and include:

1. Commercial shipping, ship construction and repair, and ship support services

2. Commercial recreation such as hotels, restaurants, shops, sportfishing, and marinas

3. Public access such as boat launching ramps, fishing piers, people-ways, and parks and open space

Further, port managers must decide whether they are to be landlords who allow the maritime industry to operate port facilities or whether the port itself will operate the facilities with its own labor force. This latter alternative does, indeed, provide public positions under the control of the port administration. These choices provide the basis for policy issues which are important to an understanding of seaports as public enterprises.

Who Governs the Seaport?

The forms of governments for seaports in the United States range across a wide spectrum. The list below includes various alternative models.[18]

1. Areawide (regional) elected board

2. Areawide board, appointed by multiple jurisdictions

[18] Not included in this list is the most independent model—a self-perpetuating board. Many nonprofit third sector organizations such as universities, research, or public interest groups use this model. Any public enterprise boards who select their own replacements without public elections or serious review by parent governments are experiencing maximum governance independence from their public.

3. Appointed board, with members appointed by a chief executive and/or legislative body of state of local government

4. Advisory board for department or state or local government

5. Department of state or local government, reporting to a a chief executive or chief administrative officer

There is no ideal model; in fact, the choice will be based on the politics in each area. Governance models will be constrained by the enabling legislation of state government. The appropriate political leaders will likely select a system which follows their tradition, which recognizes preferences of key actors, and which satisfies their particular need for independence or control.

One approach to selecting a port governance model would be to answer the question: who are the port's legitimate constituents? To do this, a dual concept will be introduced which distinguishes between those who are affected and those who are benefitted by the port.

Affected: Those in proximity to the port and who are impacted by aesthetics, environment or transportation and congestion; or those who are required to invest political energy or financial resources in a port.[19]

Benefitted: Those who directly benefit from employment and sales; or those who indirectly benefit from regional and national economic activity.

Certainly these two groups will overlap to some degree, but most individuals will take the position of one group, often at odds with the other group. Those who are affected are citizens who reside in proximity to the port and who will normally be represented by contiguous governments. Those who benefit will include employes of the port and maritime companies who might reside in proximity to the port, as well as individuals or organizations who do not live near the port but who benefit from it economically. It is assumed that those who benefit from the port, even those who reside locally, will be less concerned about the quality of life near the port than those who are affected but not benefitted.

Governance is the corporate decision of those who own the facilities and who are willing to invest in port development. In most large cities the political jurisdictions operating ports are more congruent with the group affected by the port's development and operations rather than with the broader benefitted group. The local political body or "port owners" are

---

[19] A discussion of those affected by port development is found in *Public Involvement in Maritime Facility Development* (Washington, D.C.: Maritime Transportation Research Board, National Academy of Sciences, 1979). See "Executive Summary," pp. 1-5.

also those who take the financial risk and who may contribute general tax support as the port is developed or in instances when the revenues are less than the expenditures. The business analogy for public enterprise is that ownership and risk are the traditional bases for the owner's right to govern the enterprise.[20] Voting in public enterprise is theoretically shared equally among all legitimate electors and their representatives, without regard to the size of financial risk or the benefit received.

Therefore, it is argued that a governance model should ensure representation by citizens affected by the ports in addition to businesses and employes who benefit from the port. Although customers of ports do contribute fees, they are, for the most part, not taking financial risk. In seaports, certain business and labor interests may have managed to attain power by being appointed to port boards or by directly influencing elected or appointed board members. They may, in reality, have acquired more power than their narrow economic interests justify. Citizens who live in political jurisdictions containing the port are often more concerned about issues of safety, pollution, congestion, and public access to the waterfront. They may be frustrated by having to relinquish power to those whose primary interests are financial, especially when these financial interests do not reside within their governmental boundaries. If several local governments are located close to the waterfront, or if other governments within the region feel that they are affected or that they benefit by the port, then politics may dictate a broader governance umbrella. When regional governance systems are established, then it may be expected that economic benefits will dominate the politics in lieu of the environmental impacts of development.

Port Planning

In the past, urban public works planning (including ports) was a simple matter of perceiving demands for services, gathering financial support from general taxation, intergovernment grants, and/or local debt, and then building the necessary facilities. Currently, new forms of political activism and legal opportunities are making the public works development cycle more lengthy and expensive, and are raising value conflicts which had not often surfaced in earlier development.

Seaports have not escaped the new reality that physical developments are creating impacts that are no longer acceptable to many interests. These new concerns are logical aspects of the relationship of seaport development to the surrounding community. For seaports, the key

[20] Walsh, *The Public's Business*, pp. 155-57.

planning issues can be summarized as regional port planning, environmental assessment, and coastal zone management.

Regional Port Planning

The argument for regional port planning is essentially a search for rationality in planning. Regional planning for seaports involves two basic questions.

1. Should a comprehensive regional port plan be adopted and made a requirement for individual ports?

2. Should a regional port authority be established to insure that a regional plan is implemented?

It appears that regional port planning cannot be achieved without a regional port authority. In effect, the theory is that the substance of technical planning can only be implemented with the form of a regional authority's support. The governance changes required to institute a regional authority are difficult to achieve because they curtail the competitiveness and separateness of individual ports; therefore, the question is whether rational planning can be achieved by sovereign cooperation among the individual ports.[21] It is not clear whether regional cooperation is expected to benefit the ports themselves or whether it provides an opportunity for environmental planning interests to influence development actions.

Independent ports may resist joint planning and attempt to gain a competitive advantage over other ports within their region, even though representatives of some ports meet periodically to discuss common rates. Ironically, the competition between ports is probably not determined by fee structure but by where ports are located geographically and whether or not they have reasonable transportation links. Therefore, it appears that the only financial competition is between ports within the same region. This encourages each port to avoid cooperation with neighbor ports.[22] Independence, therefore, is in conflict with regional planning, unless areawide governance exists.

[21] Jack Knecht and Stanley Euston, "Regional Port Planning and Coastal Zone Management," *Proceedings, Coastal Zone 1978*, vol. 1 (San Francisco: American Society of Civil Engineers, March 1978), pp. 37-56. The U.S. Maritime Administration has provided support for regional port studies to compare future demand and port capacity. There is little evidence, to date, that these studies have caused ports to cooperate in facility planning.

[22] This is the case in the San Francisco Bay Area with the ports of Oakland, San Francisco, Richmond, and Redwood City. In the Bay Area, the Metropolitan Transportation Commission (MTC) is attempting to encourage regional port planning in concert with the Bay Conservation and Development Commission (BCDC). MTC's power to influence port developments may only come from its ability to prevent essential transportation links to the ports. See Knecht and Euston, "Regional Port Planning," pp. 13-14.

Environmental Assessment

Seaports should not expect to escape review for conformance to federal and state environmental protection laws because they obviously have impacts on the water, air, and congestion in the community. Narrow impacts on the physical site and on adjacent marine ecology must be reviewed by means of environmental impact statements for each development project. The broader social, cultural, and economic effects of port development can be considered in more comprehensive environmental assessments of port development.[23] The value of seaport environmental review is not only environmental awareness and preservation but also an opportunity for access to the decision process by those individuals who are concerned about environmental impact. These groups are not always represented by governing boards whose primary interests are economic. One of the most frequent environmental issues arises when seaports seek disposal sites for dredging spoils. The site-specific impact on marine ecology is always great with these spoils, and serious legal debate has resulted from attempts to find ways to mitigate the negative impacts of dredging projects.[24] Of course, the inability of a port to complete dredging can threaten its maritime activities and generate anger on the part of those who are most concerned about economic activity at the port. Even though current politics may be shifting to consider more seriously a balance between the competing needs of development and ecological preservation, present legislation and regulations regarding marine wildlife and habitats create difficult legal hurdles for port development. Whereas access to the decision-making process is provided by environmental reviews, the present challenge for both the public and the port is to create ways to resolve conflicts and decrease the time needed for consensus, while retaining the due process rights of affected interests.[25]

[23] U.S. Department of Commerce, *Planning Criteria for U.S. Port Development* (Maritime Administration, Office of Ports and Intermodal Development, December 1977); see chapter 6, "Port Environmental Assessment," especially pp. 52-55.

[24] A case study on dredging spoils is found in *Public Involvement in Maritime Facility Development,* National Academy of Sciences, "Baltimore Dredging: A Dredge Site in Chesapeake Bay," pp. 115-23. See also Peri Muretta and Willard Price, "Seaport Dredging and Environmental Mitigation: The Case of Coos Bay/North Bend, Oregon," Teaching Materials for Harbor/Port Management, Institute for Marine and Coastal Studies, University of Southern California, Los Angeles, California, April 1980.

[25] One current theme in environmental management is to understand "conflict resolution" and to learn efficient methods for consensus building. See *Environmental Consensus,* a newsletter published by RESOLVE, Center for Environmental Conflict Resolution, Palo Alto, California. For an introduction to mediation, see Peri Muretta and Willard Price,

Seaports could enjoy maximum independence, but they will still be required to respond to regulatory influence from environmental agencies just as industries within the private sector are required to do. One hypothesis is that the public will have greater control over environmental issues when they have direct political access through seaport governance rather than through the legal process involved in regulation. An early commitment to environmental planning with broad participation is also suggested as useful to a resolution of environmental disputes which successfully considers the views of all parties.

Coastal Zone Management

The emergence of coastal zone management programs in the 1970s is obviously inseparable from port development. Federal and state coastal zone management programs affect seaports by drawing attention to questions of citizen participation, the concept of public access, and the planning issue of risk/hazard management.

The independence of ports and their presumed emphasis on maritime business can certainly conflict with the concept of public access to areas of the coastal zone. Public access is generally defined as allowing the public access to the water without encountering inhibiting costs, physical obstacles, or congestion. Given this concept, the greatest access occurs in areas of beaches, fishing piers, pathways, parks, and open space. Other public-oriented uses such as marinas, shops, restaurants, and hotels will restrict access to many citizens because of cost associated with the use of those facilities.[26] From the perspective of port managers, the question is whether or not public access can be provided without threatening the financial viability of the port. Limited public access, including those areas that do not produce revenue, have been successfully molded into port operations without significant impact.[27] The conflict is greatest when

"Environmental Mediation: An Alternative to Litigation," Teaching Materials for Harbor/Port Management, Institute for Marine and Coastal Studies, University of Southern California, Los Angeles, California, July 1980.

[26] A recent action of the California Coastal Commission placed public access controls on hotel developments in Marina del Rey. The commission required one hotel to set aside 15 percent of its rooms on weekends at discount rates for low-income persons, and required another to build a moderately priced hotel, a youth hostel, and inexpensive restaurants. See the *Los Angeles Times*, Part II, p. 1, August 16, 1979. A legal appeal by the developer did set aside these conditions established by the coastal commission. See *The Argonaut*, Santa Monica, April 24, 1980, p. 3.

[27] For example, the Port of Oakland has provided several sites within the port property for parks and open space near the waterfront without any serious effect on total revenues. Of course, ports that normally generate surplus public access projects may be more acceptable to port executives.

executives become adamant that the port is for maritime use and not for public access.

Some ports that are faced with declining maritime business have increased commercial recreation business, hoping to develop sufficient revenue to survive financially. Public access projects might be encouraged by coastal zone regulation, but, again, the public may have a better opportunity for influence in the matter of public access if it can reduce the amount of independence associated with planning decisions.

Another issue brought to light by coastal zone management involves hazardous cargoes and movements that take place in proximity to residential areas. Coastal zone programs are encouraging a critical look at the existing risks.[28] In risk management studies, the basic questions involve: (1) an inventory of the hazardous activity; (2) a measure of the risk of accident or natural disaster; and (3) the proximity of residents and residential developments. Naturally, these studies will lead to debate as to whether the risks are acceptable to the public. If they are not acceptable, then choices will have to be made between moving cargoes, decreasing risk, or moving land residents. Definitive risk policies will cause the port management to examine existing and new facilities, and may cause significant change in cargo operations and location, and, possibly, in the financial status of the port.

Most of these coastal zone management issues are based on a regulatory influence of state or federal agencies which are not related to the extent of port independence. However, where these issues are addressed by decision-making processes which are more open to the public input, then the amount of planning independence of the port enterprise is clearly diminished. This is consistent with the normative view that planning and governance is expected to be less autonomous.

Is There a Need for an Expanded Federal Port Policy?

If the federal government increases its role in seaport development, does that mean that ports will be less independent? Certainly port executives will resist increased regulation and requirements for port development, but will they also reject financial support from the federal government? What port executives appear to want is increased grants for capital projects and reimbursement for costs mandated by other governments. But they do not want regulations that will increase costs or create delays in development. Ports are no different from other local

---

[28] Jeffrey A. Finn, "Energy and the Coastal Zone: A Question of Risk," the theme issue in a special edition of *Coastal Zone Management Journal* 7 (1980).

programs in this regard. What makes the ports different is that they may not be able to argue that they need federal grants.

At the present time there is a modest federal contribution to the capital and operating costs of sea ports, but that contribution is not a large share of the costs, nor is the share likely to increase as the federal government moves toward a balanced budget. Recently, the U.S. Comptroller General proposed a range of options for the Congress to consider regarding seaports, from existing policies to federal underwriting of capital investments and operative deficits.[29] These bold initiatives seem inconsistent with a reduced federal budget. If any change is likely, it would be toward higher port contribution to the existing federal programs.

The Department of Commerce, through the Maritime Administration has a program to support research and planning in ports. The Coast Guard provides valuable ship safety services for ports in much the same way as the Federal Aviation Agency supports airports. The objective of the Coast Guard is to ensure the public safety in and near the harbor, something that the port may not choose to do on its own. The most significant federal program is the long-standing program of the U.S. Army Corps of Engineers to conduct dredging projects in most harbors. These projects are critical for most ports to maintain existing depths and to respond to the new generation of ships which require deeper channels. One question is whether or not ports are prepared to assume an increasing share of the cost of dredging. The dredging program has traditionally been a "pork barrel" program, but the Congress must now consider increased cost sharing. Why should the ports resist paying all or part of the dredging cost, since they can pass on the costs to the customers who are benefitting from the available depth? If seaport enterprises are to be independent, the federal support cannot be defended, especially if no other public interest can be identified. One argument which can be made is that port investment is simply one avenue of economic stimulus—a proper objective for the federal level.

The Economic Development Administration (EDA) of the Department of Commerce continues to make funds available to ports capital projects, although EDA's objective is not to aid the ports but to invest in projects

[29] U.S. Comptroller General, *American Seaports—Changes Affecting Operation and Development*, CED-80-8 (Washington, D.C., 1980), pp. 22-26. For a discussion of the federal involvement in seaports, see Henry S. Marcus et al., *Federal Port Policy in the United States* (Cambridge, Mass.: MIT Press, 1976), pp. 301-8; and Marcus, "The Role of the U.S. Federal Government in Port Planning and Development," in Willard Price, *Is There a Federal Port Policy?* Teaching Materials in Harbor/Port Management, Institute for Marine and Coastal Studies, University of Southern California, Los Angeles, California, March 1980, pp. 27-31.

that will increase short-term and long-term employment. Often the ports with the greatest need for capital development are those with healthy economic activity, and they may not have the unemployment rates to justify EDA's grants. In smaller ports or in economically deprived regions, seaports might be operated with a more comprehensive economic development purpose, seeking to enhance employment and economic activity as a principal objective by any means. Maritime activity might be only one function of the enterprise.

Therefore, is it not clear how an expanded federal role in seaports could be justified? Most ports which need to expand are successful and have sufficient revenues available to finance new capital facilities and operation. Without an established objective, continued federal involvement may not be expected, even though subsidy is defended in other enterprises such as mass transit, water resources, housing, and energy.

What does all this mean for independence? In the past, many enterprises were established which recognized government subsidy as a necessity, some for an initial period and others more or less permanently.[30] Now there appears to be an inevitable trend toward self-sufficiency of public enterprises whenever possible, while there may be no such solution for mass transit, water resources, or public housing; seaports may, therefore, be destined to be financially independent from the federal government.

Are Public Enterprise Finances Really Independent?

One common expectation is that public enterprises ought to be financially independent. It is argued that public enterprise managers cannot perform effectively without financial autonomy. In the subsequent discussion, we will show that while public enterprise may operate under separate accounting systems, several other aspects of financial management policies will not be free of political influence.

If a public enterprise is to maintain any separate identity and accountability, then a separate accounting system is mandatory. If revenues and expenditures are comingled with other programs, there is no basis for judging financial performance or for creating a public enterprise. But a separate operating statement and balance sheet do not necessarily mean that people outside the public enterprise cannot affect accounting entirely. However, decisions to influence accounting reports

[30] Sherwood, "Small Group Discussion," p. 35.

cannot be made unless such documents are public information, indicating the choices which are available for accounting practice.[31]

Determining fees for services in seaports is one of the most technically difficult tasks in public enterprise finance. The level of fees will determine financial health at the same time it affects demand for services. To ensure adequate recovery of expenses and, at least, a balanced budget, clearly a rational fee policy must be calculated. Capital and operating costs, as well as appropriate overhead, must be compared with the estimated business before a rate can be determined. Any changes in expected costs or sales will change the revenue flow. But this rational method represents only one view. Other approaches are possible:

1. Set fees on the basis of "what the market will bear" (public and private enterprises find this method simpler and possibly more profitable)

2. Set fees in accordance with agreement of competing ports (in effect, this eliminates competition by price)

3. Set fees under political influence from those who approve rate structures (certainly some interest groups are able to marginally affect fees to their advantage)

The defense of each of these methods is fairly obvious, and only a normative position can choose the correct approach. As public enterprises, ports cannot expect to keep their rate determination method from an open review and criticism. Only private organizations are able to defend secret pricing policies under the guise of competition.

In any private organization, debt is necessary to allow capital expansion. It is not considered a poor financial policy to be in debt, provided the enterprise keeps the ability to repay its obligation and to remain financially viable. Accounting standards do set criteria for the amount of debt as a percentage of income or assets in order to provide some cushion against uncertainty in revenues. The availability of debt financing is critical to a seaport's ability to respond to demand, and those who can authorize debt retain the ability to stop expansion or improvements. One issue is whether ports, and public enterprise in general, are able to sell tax-free revenue bonds without public (owner) approval.[32] The question of independence depends on whether the public allows the enterprise to extend its debt freely. Control over debt carries with it the power to affect facilities and prices and to affect the risk of deficits.

[31] Walsh, *The Public's Business*, p. 345.
[32] Ibid., pp. 55-83.

The net result of the revenue and expenditures in any year is the most controversial issue in seaport financial management. Deficits are not normally planned in any public corporate entity (cities, states, or public enterprises). But in the early periods of port development, subsidies are often necessary to prevent deficits, because massive capital and operating costs cannot be recovered until business is developed. Local governments may have contributed funds directly or serviced general obligation bonds from general tax dollars. This can be viewed as the startup costs or seed money necessary for local investment in the port enterprise. Once the port has become established, it is expected that surpluses will result and that possible repayment will be made to the general fund.

For the most part, ports are financially successful, generating surpluses after all debt obligations have been paid. In the private sector, this might be called gross profit before taxation and net profit before dividends. After dividends, what remains is retained earnings. Public enterprises do not necessarily pay taxes or dividends, but they may, in effect, pay something very similar. There are some policy options for the disposition of surplus funds:

1. *Depreciation* is a traditional accounting technique and should be reserved by ports before surplus is declared. However, depreciation accounts do encourage the replacement of aging facilities and limit other financial choices.

2. *Overhead* is a delicate subject in public enterprise. Besides the obvious overhead expenses of the enterprise management, staff, and other general services, there are services which may be rendered by other governments for which no contract is in effect. In those situations, governments that have influence over enterprises may demand additional overhead payments.[33] The issue is whether or not those payments represent defensible costs, and if so, should they be shifted to formal contracts. It is possible that parent governments will seek payments which are higher than can be justified, in essence, charging enterprises a fee "in lieu of taxation."[34]

3. *Capital Reserve* accounts can be established to set aside excess

[33] California's Proposition 13 has caused some cities with enterprise operations to increase their demand for overhead contributions from the enterprise for administrative, legal, public safety, or other services. Maybe enterprises have received free services when local governments were not concerned about cost accounting.
[34] The concept of "in lieu of taxation" payments by public enterprises has been established in New York. See State Charter Revision Commission for New York City, *Public Benefit Corporations in New York City,* January 1975, p. 18 and pp. 38-43.

revenues as a reserve against future capital needs without any specific project in mind. This conservative fiscal tactic encourages management to build additional facilities and extend their risk of over-capitalization (overcapacity).

4. *Public Dividends* can be paid with excess funds in the same way that private companies declare dividends. Since stockholders receive those monies as income, it should be feasible for a parent government to receive income from the enterprise and to spend it for any other public function.[35] The return goes to the owner and risk taker. Who would bail out the enterprise if deficits occurred?

Favorable budget conditions will produce surpluses which seaports need for modernization and expansion of maritime facilities. Ports can build on the basis of available funds, but should they not test the public's preference for additional capital investment? What the public needs to know is whether development is creating overcapacity. The port needs to openly defend excess capacity in terms of flexibility, competition or decreased ship delays.[36]

Port expansion does not have to be limited to maritime facilities; surplus funds may be better used for commercial recreation or public access, although they generate less fees and may not completely pay their capital and operating costs. It is true that surplus revenues from shipping customers will support other port uses, but why not argue about the cost of using the waterfront and denying access to the public? Another position is that surplus is a dividend; it is paid to the owners of the port and they may invest in alternative uses of the waterfront.

If funds become available to general governments, then port surpluses support other public programs not able to raise revenues. Certainly, water-related public works or transportation systems are logical extensions of the port enterprise. Where and when does it become unrelated to port activities? The Port Authority of New York, along with other ports, has found commercial property development to be a desirable output for surplus revenues. Every port wants a "World Trade Center," and many ports can afford to finance such superstructures.[37] Are there better uses for monies?

[35] Ports in California argue that they are restricted by state statutes from using excess funds to support other governments. Contrasting California is the Port of Baltimore, owned by the Maryland Transportation Department, which normally co-mingles surplus port funds with other transportation revenues.

[36] Marcus, *Federal Port Policy*, pp. 206-8.

[37] Leonard I. Ruchleman, *The World Trade Center: Politics and Policies of Skyscraper Development* (Syracuse, N.Y.: Syracuse University Press, 1977).

One more legitimate use of surplus dollars is as a contingency against the economic uncertainty which ports can face. Technology or competition may eliminate business, and a cushion can allow an enterprise to shift directions or to weather economic cycles. If accounting information on contingencies is withheld, if the port is too independent financially, then decisions might be made on a narrow policy basis.[38]

Concluding Comments

This chapter attempted to draw attention to the issues involved in seaport mangement through a focus on the concept of public enterprise. Public enterprise should become a more significant part of the practice of public administration, given scarce resources and increased public scrutiny of government. If the public's interest is in more independent delivery of essential public services, then the enterprise approach can be considered for an increasing number of functions that remain owned by the public sector.

Just as important to the maturation of the enterprise model is the attention of academics and research programs in schools of public policy and management. Hopefully, additional research support will be made available from foundations, federal agencies, and the port and maritime industries. It was argued that public enterprise is more public than enterprise and that independence is not guaranteed in public ownership. The conclusion is that the field of public administration should take the academic lead. Public enterprise research will need to take into account the perspective of business administration as well as of political science, international relations, economics, and urban planning. Seaport management will also need to specifically interface with the fields of civil engineering/public works, marine policy, water resources, transportation, and coastal zone/environmental management. Public enterprise research represents another area for multidisciplinary study which cannot mature without some consensus among these several disciplines regarding research approaches.

Many individuals in society are criticizing public management for presumed inefficiencies and at the same time demanding more openness and participation in the significant enterprises of both the public and the private sector. The academic and professional opportunity to address these concepts is obvious.

[38] State Charter Revision Commission, *Public Benefit Corporations*, pp. 31-33.

CHAPTER 12
TOWARD A POSITIVE MODEL OF
FISHERY MANAGEMENT DECISION-MAKING

TIMOTHY M. HENNESSEY*

First we had the Russians, and now we have you; I prefer the Russians, there were fewer meetings.[1]

Introduction

Natural resources with common property characteristics are particularly vulnerable to unlimited and potentially harmful exploitation by individual entrepreneurs. Consequently, a powerful case can be made that the use of such resources should be regulated by the government. We have followed such an approach in the United States for many natural resources. This paper concerns the management of one of these—fisheries, and the role the federal and state governments play in managing this resource *via* the Fisheries Conservation and Management Act of 1976.

The Fisheries Conservation and Management Act established a national management program to regulate fishing activities within the 200-mile fishery conservation zone. The management program and the institutional arrangements to implement it are quite complex and represent particularly interesting examples of intergovernmental relationships among the states, regions, and the federal government.

The objective of the Act is clearly specified as the attainment of optimal fishery yield. This is basically maximum sustainable biological yield adjusted for economic, social, and ecological factors. The attainment of this objective, however, is subjected to a series of powerful constraints. Principal among these is that decisions are made by eight regional fishery management councils that necessarily operate under conditions of imperfect information, uncertainty, and risk. Moreover, the fishery management plans and fishing quotas result from bargaining within the councils and interactions between the councils and the National Marine

* Timothy M. Hennessey is Professor of Political Science and Director of the Master of Public Administration Program at the University of Rhode Island. Some of the material in this chapter was originally presented to a conference on Formulating Marine Policy: Limitations to Rational Decision Making, Center for Ocean Management Studies, University of Rhode Island, June 19-21, 1978. The conference was sponsored by the Center and the Office of Sea Grant of the National Oceanic and Atmospheric Administration. Permission to use this material is gratefully acknowledged.

[1] Galilee, Rhode Island Fisherman to a Representative of the National Marine Fisheries Service, 1979.

Fisheries Service. The latter must not only approve council decisions but must also provide the councils with the technical information necessary for them to make the decisions. There is also considerable bargaining and interaction among the regional councils themselves insofar as they have adjoining fisheries. Finally, there is interaction between the regional councils, the Department of Commerce and the Department of State, insofar as foreign policy issues related to fishery management are concerned.

Thus, the decision-making system for the Fisheries Conservation and Management Act (FCMA) has what seems to be a relatively clear-cut set of objectives, at least in a technical sense. Attaining these objectives, however, is to a great extent a function of informational and institutional factors that are not technical. The implementation of the Act takes place within an institutional system the dynamics of which are not understood. This paper attempts to enhance understanding of this complex process by specifying a set of empirically testable hypotheses by which to study comparatively the dynamic interactions among technical issues of optimal yield, institutional structures, and information as these relate to regional management council decision-making. Given these constraints and the complexities of interaction and bargaining which characterize council decision-making, the usual decision-making models (what Lindblom terms synoptic) are inappropriate to an understanding of council decision-making since they rely on high-level information and hierarchical decision structures. I argue, instead, for an approach that employs Lindblom's concept of partisan mutual adjustment processes and disjointed-incremental decision patterns.[2] I employ the latter approach to begin to develop a positive model of the council decision-making process.

The Fishery Conservation and Management Act: Objectives and Structure

Responsibility for administering the national fishery management program resides in the Department of Commerce, or more specifically, the National Oceanic and Atmospheric Administration's National Marine Fisheries Service. Primary responsibility for the preparation of the management plan is located with eight regional fishery management councils. The eight are as follows:

[2] Charles E. Lindblom, *The Intelligence of Democracy: Decision Making Through Mutual Adjustment* (New York: Free Press, 1965), *passim*; Charles E. Lindblom and David Braybrooke, *A Strategy of Decision* (New York: Free Press, 1963), pp. 21-147.

1. New England Council, consisting of the states of Maine, New Hampshire, Massachusetts, Rhode Island, and Connecticut; with 17 voting members

2. Mid-Atlantic Council, consisting of the states of New York, New Jersey, Delaware, Pennsylvania, Maryland, and Virginia; with 19 voting members

3. South Atlantic Council, consisting of the states of North Carolina, South Carolina, Georgia, and Florida; with 13 voting members

4. Caribbean Council, consisting of the Virgin Islands and Puerto Rico; with 7 voting members

5. Gulf Council, consisting of the states of Texas, Louisiana, Mississippi, Alabama, and Florida; with 17 voting members

6. Pacific Council, consisting of the states of California, Oregon, Washington, and Idaho; with 13 voting members (Idaho is included because Pacific salmon migrate up the Columbia River and its tributaries into Idaho, where major spawning areas are located)

7. North Pacific Council, consisting of the states of Alaska, Washington, and Oregon; with 11 voting members

8. Western Pacific Council, consisting of the state of Hawaii, and the territories of American Samoa and Guam; with 11 voting members.

Decisions are reached by voting. The voting members on each council are representatives from its member states and from the federal government. A council consists of the principal state official with marine fishery management responsibility as selected by the governor, the regional director of the National Marine Fisheries Service, and a remaining number of members equal to twice the number of member states appointed by the Secretary of Commerce from a list submitted by the governors of the member states. The latter are "qualified individuals" knowledgeable or experienced in management, conservation, or recreational and commerce harvesting. There are also nonvoting members who include the regional director of the U.S. Fish and Wildlife Service, the commander of the appropriate Coast Guard district, the director of the Marine Fisheries Commission, and one representative of the Department of State. These eight councils are responsible for developing fishery management plans and amendments. They review and revise assessments of optimum yield and allowable foreign fishing. They are also required to hold public hearings on management plans and to establish scientific and statistics committees which give technical advice to the councils concerning optimum yield and allowable catch.

The latter task is demanding, since each plan must contain all conservation and management measures which are accompanied by a complete description of the fishery, including the number of vessels involved, gear type, revenues, and recreational use. Finally, the plan must contain an assessment of present and future projects regarding the condition of the fishery, including both maximum sustainable yield and optimum yield. These plans must also be consistent with the following seven national standards:

1. Conservation and management measures must prevent overfishing but achieve optimal yield for each fishery.

2. These measures must be based on the best scientific information available.

3. To the extent practicable, an individual stock of fish should be managed as a unit through its range.

4. Conservation and management measures should not discriminate between residents of different states.

5. The measures should promote efficiency in the use of fishery resources.

6. The measures shall take into account variations among fisheries and fisheries resources and catches.

7. Where practicable, the measures should minimize costs and avoid unnecessary duplication.[3]

It should be clear that a sophisticated data system and decision process are required in order to meet these standards. It is to these requirements that we now turn.

Relevant Analytics and Informational Limitations

Prior to passage of the Fishery Conservation and Management Act of 1976, the most common approach to fishery management was based on the concept of maximum sustainable biological yield, sometimes modified

---

[3] Fisheries Conservation and Management Act of 1976.

by economic considerations. Research has shown that when virgin fishing stocks are fished, total catch will be high and the rate of catch will be high initially. However, as more vessels fish the stock, its size decreases and the net growth rate approaches its peak. The level at which this peak or high level of total catch is reached is called maximum sustainable yield.

If a stock is fished at the level of maximum sustainable yield, the total catch will be maximized, but individual catch rates will be lower and lower as the maximum sustainable yield is approached. Once the fisherman's catch rate is low he will be near the break-even point or will actually lose money. That is, he will be increasing his fishing effort to obtain the same catch of fish.

Therefore, if the sole purpose of fishing is to maximize poundage of fish caught, regardless of cost, then maximum sustainable yield is the best approach. However, if the purpose is to maximize earning from effort (which is clearly the case for fishermen), fishing must be at a yield level where the economic costs of fishing are taken into account. It should be obvious that the more the stock is fished, the faster maximum sustainable yield is reached, and the more it costs to fish.

The Fishery Conservation and Management Act takes into account the economic optimum, which is the point where net returns from the total fishery are highest before the maximum sustainable yield level is reached. In other words, one way to approach catch quotas in light of maximum sustainable yield adjusted for economic considerations is to use economic catch rates, within the biological limits. Figure 1 illustrates this approach.[4]

Note the distance between the dollar value of the catch and the cost line at *a* and the maximum sustainable yield level *b*. Even when the economic optimum is calculated, difficult as this may be, this is not adequate to get optimal yield. The Fisheries Conservation and Management Act has its objective optimum yield defined as the amount of fish which (a) will provide the greatest overall benefit to the nation, with particular reference to food production and recreational opportunities; and (b) which is prescribed as such on the basis of the maximum sustainable yield from each fishery as modified by any relevant economic, social, or ecological factors.

This approach has enormous analytic and informational requirements. Not only must one have sound biological and catch data but economic, social, and ecological information as well. Moreover, a methodology is required by which maximum sustainable yield can be modified mathematically to reflect economic, social, and ecological factors directly.

---

[4] "Fisheries Science, How and Why It Works for Marine Fisheries," *Fisherman's Information Bulletin*, 78-3B, August 1978, Government of Canada, pp. 3-4.

Figure 1

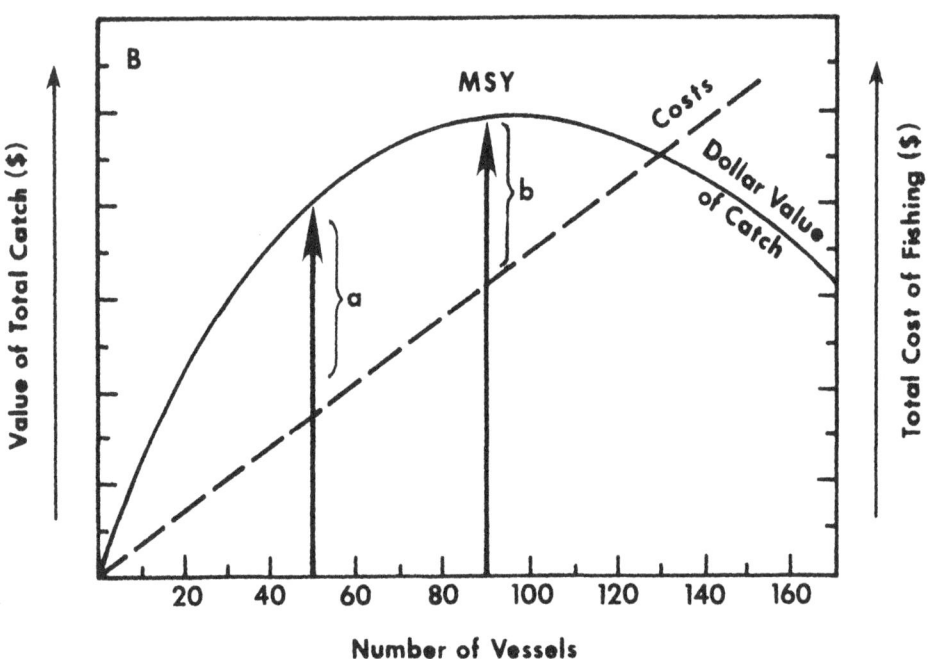

Needless to say, one expects existing information levels for these factors, to be extremely low, and even if adequate information were available the analytics are not adequate at this time in terms of which such variables can be directly incorporated into the calculations.

Such analytic and informational limitations make it extremely difficult and risky for the councils to make decisions in the face of such imperfect information. Even if the councils had better information, they would still face time and resource constraints which would force them to seek simplifying decision-rules to reduce the costs associated with continued information search. I shall attempt here to suggest some testable hypotheses which flow directly from these circumstances.

Beyond informational requirements, the decision-making process of the councils is further complicated by the structure of the fisheries management councils referred to above and their relationship to the federal government. Under the provisions of the Fisheries Conservation and Management Act, management plans are prepared by Regional Fisheries Management Councils, but they are implemented by the federal government. The National Marine Fisheries Service also plays a technical support role to the councils by providing operating funds and data on which domestic quota decisions are made. Yet the Service also performs a role for the Secretary of Commerce in advising on which of the plans is basically sound. This dual role of serving the councils as well as the Secretary of Commerce is often seen as a cause of some confusion and a potential source of conflict of interest between the federal government and the states.

In sum, the optimizing approach specified in the Fisheries Conservation Management Act not only has severe analytic and informational limitations, but the decision-making process is further complicated by the structural and procedural specifications of the act. There are a multiplicity of decision points involved in the process of implementing the act, and this tends to increase the decision-making costs associated with any particular decision or set of decisions. Given the conditions of imperfect information, uncertainty and risk, and the multiplicity of decision points and the associated decision costs, it should be emphasized that the usual decision-making models are inappropriate since they require a high level of information and low decision-making costs. Moreover, the utilization of the usual decision models is not only inappropriate but potentially misleading, in that such models can lead us to have higher expectations for management council decisions than the actual outcomes approximate. This could erode public confidence in the management process.

If we are to begin to understand the decision-making process in fisheries management councils, we must employ models which explicitly recognize the limitations identified above. Using such models, I shall formulate empirically testable hypotheses regarding the comparative decision-making behavior of the management councils. Unfortunately, this has not been done previously and the literature which does exist on the councils is largely descriptive.[5]

Decision-Making Models: Synoptic vs. Strategic

Our understanding of the decision-making process in fishery management councils is dependent on the type of decision model employed. I argue for the development of a positive model which recognizes the empirical conditions and constraints appropriate to the decision environment of the councils. Following Charles Lindblom and Herbert Simon, I shall employ a strategic or bounded rationality approach. This approach will be contrasted with the usual synoptic decision model.

The synoptic model is an optimizing approach which selects that course of action with the highest payoff. This requires estimating the value of each alternative in terms of benefits and costs. According to Irvin Janis, "quality decisions" using the synoptic approach should be achieved as follows. A decision-maker would:

1. Canvass a wide range of alternative courses of action.

2. Survey the full range of objectives to be fulfilled and the values implicated by each.

3. Carefully weigh whatever is known about the costs and risks of negative consequences as well as positive consequences which could flow from each alternative.

4. Intensively search for new information relevant to further evaluation of the alternatives.

5. Correctly assimilate and take account of new information or expert judgment even when the information or judgment does not support the preferred course of action.

[5] H. Gary Knight, "Management Procedures in the U.S. Fishery Conservation Zone," *Marine Policy*, January 1978, pp. 22-29; Alan J. Wyner, John E. Moore, and Bilana Cicin-Sai, "Politics and Management of the California Abalone Fishery," *Marine Policy*, January 1978, pp. 10-20; and John E. Kelly, "The Fishery Conservation and Management Act of 1976," *Marine Policy*, January 1978, pp. 30-36.

6. Reexamine the positive and negative consequences of all known alternatives, including all those originally regarded as unacceptable, before making a final choice.

7. Make detailed provisions for implementing or executing the chosen course of action, with special attention to contingency plans that might be required if various known risks were to materialize.[6]

This approach is obviously an ideal which few decision-makers can attain. There is a variety of explanations why such a decision-making ideal is beyond the reach of administrative man. First, this approach ignores the fact that since information is costly to acquire, the search for it is often not worth the cost. And since comprehensiveness is required by synoptic models, and such comprehensiveness requires high levels of information, it follows that decision-makers will be forced to forego comprehensiveness. Instead of the synoptic ideal, one might more realistically view the decision-maker as laboring under the powerful constraints of incomplete information and the uncertainty associated with it. The decision-maker is also very sensitive to the opportunity costs associated with the analysis itself; the more complex the analysis, the higher the opportunity costs. In sum, the synoptic approach does not reflect the types of cost constraints facing the decision-maker and, therefore, is not an acceptable approach to the development of a positive model of decision-making.

There is, however, a body of theory which can be subsumed under the label *bounded rationality* which contains a set of concepts for understanding decision-making in public organizations under constraint. Focusing on such decision-making problems, Charles Lindblom sees decision-making as fundamentally a problem of *adaptations* to constraints.[7] For him, adaptation is the key to survival in the face of complexity. He argues that, at a minimum, one should design a decision-making system which takes into account:

> (1) man's limited problem-solving capability, (2) the inadequacy of information and knowledge, (3) the costliness of analysis, (4) the openness of the system of variables, (5) the analyst's need for strategic sequences of analytic moves, (6) the analyst's inability to construct a complete rational-deductive system, and (7) the diversity of forms in which policy problems arise.[8]

[6] Irving Janis, *Decision Making*, (New York: Free Press, 1977), p. 11.
[7] Lindblom, *A Strategy of Decision*, pp. 47-57.
[8] Ibid., p. 113.

Lindblom argues that in most public organizations analysis can be characterized as "disjointed—incrementalism." That is, it is "incremental, restricted (both with respect to variety of alternatives and to the variety of consequences of each alternative), means- oriented, reconstructive, serial, remedial, and fragmented."[9] He sees the decision-maker as concerned with the margins at which situations might be changed from the status quo. He calls this *margin-dependent choice.* Only those policies are considered which differ *incrementally* from the status quo. Moreover, the examination of policies proceeds through comparative analysis of no more than the marginal or incremental differences in the consequent social states rather than through an attempt at more comprehensive analysis.

While the conventional view of problem solving is that means are adjusted to ends in that policies are sought that attain certain objectives, Lindblom argues that the reverse is more likely to be the case. For him, *the proximate ends of public policy are governed by the means available.*[10]

He sees the analyst choosing as relevant objectives only those worth considering in view of the means actually at hand or likely to become available. He automatically incorporates considerations of the costliness of achieving the objective into his marginal comparison for an examination of incremental differences in value consequences of various means. This tells him what price in terms of one value he is obtaining for an increment of another. And while he contemplates means, he simultaneously contemplates objectives unlike the synoptic analyst who ideally must stabilize his objectives and then select means.[11]

It is also to be expected that stable, long-term aspirations will not appear as dominant values for policy-makers. As Lindblom notes:

> The characteristics of the strategy support and encourage the analyst to identify situations or ills from which to move *away* rather than *goals* toward which to move.[12]

In this sense, *policy is fundamentally remedial.*

Those who employ this strategy will behave approximately as follows:

> The analyst makes an incremental move in the desired direction without taking upon himself the difficulties of finding a solution. He disregards many other possible moves because they are too costly (in time, energy and money) to examine. For the move he makes he does not trouble to find out what all its consequences are. If his move

---

[9] Ibid., p. 112.
[10] Ibid., p. 93.
[11] Ibid., p. 94.
[12] Ibid., p. 102.

fails or is attended by unanticipated adverse consequences he assumes that one's next move will take care of resulting problems. If his policy-making is remedial and serial his assumptions are usually correct.[13]

This view of problem solving as successive approximation is a practical and sophisticated adaptation to the impossibility of attaining the synoptic ideal.

Another approach to finding alternatives which explicitly recognize constraints is a *satisficing* model of public policy. Herbert Simon uses this approach to move beyond optimizing computational techniques to deal with *constrained choice as it yields satisfactory performance.*[14] He refers to "figures of merit" which permit comparison between designs in terms of better or worse but which seldom provide a judgment of best. Simon argues that no one in his right mind will satisfice if he can equally well optimize: no one will settle for good or better if he can get best. But this is not the way the problems present themselves in the public arena.

> An earmark of all these situations where we satisfice for inability to optimize is that, although the set of available alternatives is "given" in an abstract sense, it is not given in the only sense that is practically relevant. We cannot, within practicable computational limits, generate all the admissible alternatives and compare their respective merits. Nor can we recognize the best alternative, even if we were fortunate enough to generate it early, until we have seen them all. *We satisfice by looking for alternatives in such a way that we can generally find an acceptable one after only moderate search.*[15]

To this point we have spoken largely in terms of single decision-making units and have not considered the applicability of the two alternative models to complex public organizations. But the organizational setting is critical, because such organizations develop routines of behavior which reinforce incrementalism and serial behavior.

In this regard the work of John Steinbruner offers a view of routines of organizational behavior which complements Lindblom.[16] Both base their arguments on the fundamental assumption of man's incapability to deal in any comprehensive manner with the complex demands of the

[13] Ibid., p. 123.

[14] Herbert Simon, *The Sciences of the Artificial* (Cambridge, Mass.: MIT Press, 1969), pp. 63-64.

[15] Ibid., pp. 62-64. See also, Herbert Simon, "Rational Decision Making in Business Organizations," *American Economic Review*, Fall 1979; and Herbert Simon and James March, *Organizations* (New York: Wiley & Sons, 1958).

[16] John Steinbruner, *The Cybernetic Theory of Decision* (Princeton: Princeton University Press, 1974), p. 66.

environment. Steinbruner, like Lindblom, argues that the key to his cybernetic approach is *the control of uncertainty*. The decision-maker does this by focusing on a few incoming variables which eliminate any serious calculations of probable outcomes. As he observes,

> The cybernetic decision-maker is sensitive to information only if it enters through an established highly focused feedback channel and hence many factors which in fact affect outcomes have no effect in his decision process.[17]

Steinbruner argues that greater complexity entails greater variety, and that under conditions of great complexity the decision maker must have a more elaborate response repertory. Internal simplicity can be preserved under complexity if the number of decision makers concerned with the problem is increased. Each decision-maker can then focus on some limited dimension of the complex problem. Lindblom has recommended a similar approach in his theory of partisan mutual adjustment, a system in which coordination is maximized through a multiplicity of strategic problem solvers.[18] This can be contrasted with the synoptic approach to the problem of complexity which is to array the organization in a single-centered, hierarchical fashion with information flowing upward to the decision maker.

Lindblom then sees coordination in complex public agencies as taking place via interactions between strategic problem-savers. Since synoptic problem solving is impossible, a multiplicity of decision-makers cope with the inevitability of error and omission which are endemic to complex problem solving. It is through such multiplicity that

> decision makers mop up the adverse consequences of each others inevitably imperfect decisions. Multiple decision makers will, in addition, compellingly call to others' attention aspects of the problem they themselves cannot analyze. Moreover, just as the single decision maker will sometimes anticipate adverse consequences that he must nevertheless treat as separate problems for fear of making his existing problem unmanageable, so also other decision makers can anticipate what he either cannot anticipate or cannot attend to. The relationship of strategic problem solving to partisan mutual adjustment now begins to be clear.[19]

Given the distinction between synoptic and strategic decision-making models and the respective advantages of the latter over the former, let us turn for application of our argument to the Fisheries Conservation and

---

[17] Ibid., p. 67.
[18] Lindblom, *The Intelligence of Democracy*, p. 157.
[19] Ibid., pp. 151-52.

Management Act of 1976. In what follows I shall hypothesize that the Act uses strategic interactions among the federal government and the regional councils, as well as interactions within and among the councils, in order to minimize inherent information limitations. This multiplicity of decision points maximizes a diversity of viewpoints and preferences. In this sense, the management system substitutes social interaction for comprehensive analysis, and uses conflict as a mechanism of discovery.[20]

Toward a Positive Model of Council Decision-Making

The strategic approach allows us to investigate the complex setting of management council decision-making. Initially, I hypothesize that the full range of variables associated with optimum yield will simply not be considered by the councils. Some councils will give more attention to a broad range of factors than others, but no council will be as comprehensive in its treatment of these variables as the traditional decision models require, or the goals of the Fisheries Conservation and Management Act specify.

This follows from the costs to the decision-maker of acquiring and utilizing information as well as the lack of an adequate data base. But even if high levels of information were readily available, decision-makers would still seek to reduce their decision costs by ignoring or discounting certain information. In the case of the Regional Fisheries Management Councils, we would expect them to reduce the range of information which they require. Following Simon[21] we expect them to use only that amount of information that permits them to *satisfice*, with respect to some preferred course of strategic action. All other information will be ignored. The observations of other scholars reinforce such a hypothesis. Lindblom, for example, argues that decision-makers, given the limitations of time and costs of analysis, will confine themselves to only a limited number of factors in making policy. This narrowing procedure allows them to concentrate on these few while ignoring others. As he notes, "to omit is to make manageable."[22] Precisely what types and levels of information would *satisfice* for different council decision-making strategies is one of the most important general avenues of inquiry in understanding council behavior.

The relevance of these observations for decision-making in the Regional Fisheries Management Councils and for the overall goal of

---

[20] Charles E. Lindblom, *Politics and Markets* (New York: Basic Books, 1977), pp. 253-56.

[21] Simon, *Sciences of the Artificial*.

[22] Lindblom, *Strategy of Decision*, p. 81.

optimum yield would seem to be as follows. Let us assume that the primary, although not exclusive, consideration for the councils is the availability of biological data in terms of which to calculate maximum sustainable yield for management plans and quota determination. The Regional Fisheries Management Councils have fisheries under their jurisdiction which vary considerably with respect to the degree of their complexity. *By complexity we mean the number and variety of different species of fish as well as the characteristics of the fishing industry in terms of vessel size and gear types.* The use of the term "complex" will denote variations along this dimension for both the fishery and fishing industry. In this sense, there is considerable variance across the eight councils. For example, the New England Council has a complex fishery while the Gulf Council has a far less complex one. Variance in the complexity of the fishery creates an accompanying variance in the informational costs facing the councils. This line of argument suggests the hypothesis that: *The more complex the fishery, the greater the informational needs and decision-making costs for the councils.*

Since we assume that the fishery management council decision-makers, like all others, seek to minimize the costs associated with decision-making, we expect them to engage in strategies that permit concentration on that information which is the most salient for immediate decisional needs while giving less attention to other less salient information.

We can conceptualize this cost-reducing process by considering that the *range of issues* facing the councils is at least of the following four types: biological, economic, social, and ecological. All of these are required for calculating optimum yield, albeit biological factors are clearly the most central for determining maximum sustainable yield. Conceptualizing issue-range in this manner, I hypothesize that: *All other things being equal, the more complex the fishery, the narrower the range of informational concerns the management council can have.* Conversely, the less complex the fishery, the broader the range of informational concerns the council can have.

Councils with more complex fisheries will seek to "make things manageable," in Lindblom's sense, by devoting the greatest proportion of their scarce time and the time of their staff to the acquisition, organization, and interpretation of the basic biological data required to ascertain maximum sustainable yield. This will be followed by economic data. Other factors will receive the least treatment since data on social and ecological matters are not currently available, can be acquired only at considerable cost, and are still difficult to utilize in modifying maximum sustainable yield. Thus, since even the basic task of calculating maximum sustainable yield is demanding, biological data will be the most salient concern; economic, social, and ecological concerns will be pushed to the margins in the formal information acquisition effort. This in no sense means that councils see nonbiological concerns as unimportant but rather that efforts to gather information on social and ecological factors are simply not feasible, given time and cost constraints.

This does not mean, however, that because of this narrowing of the range in councils with complex fisheries, information on economic, social, and ecological questions will not be forthcoming. Indeed, it is precisely because the council does not have such information that mechanisms other than *formal* information gathering must be found. If such mechanisms cannot be discovered and utilized in the decision-making process, councils with complex fisheries will not be able to approach optimum yield which requires that maximum sustainable yield be adjusted in light of economic, social, and ecological factors. If councils cannot make such adjustments, where appropriate, they will have failed to achieve the goals of the Fisheries Conservation and Management Act. Fortunately, such a mechanism is inherent in the institutional structure and dynamics of the councils themselves; namely, the process of partisan mutual adjustment.

Since there are numerous decision points specified in the Act, I expect the dynamics of council interactions and the triadic relationship between the Secretary of Commerce, the Regional Fisheries Management Councils, and the National Marine Fisheries Service to generate a diversity of viewpoints and preferences. Under such a system there is a higher probability that negative impacts of particular courses of council action will be brought to the surface than if the system were centrally coordinated and employed synoptic operational criteria.

The Fisheries Conservation and Management Act specifies an organizational structure that brings forth partisan mutual adjustment, and social interaction substitutes for comprehensive analysis. The interaction is played out as a struggle between competing interests. Conflict is inevitably high in the management councils, but this can be beneficial to the overall objectives by bringing out facts which would not otherwise be discovered, given the necessary limits of time and analysis. As Lindblom observes, "In the mutual-partisan-adjustment process, conflict can be used as a mechanism of discovery."[23] I hypothesize then that *the more complex the fishery the greater the informational needs, and the more the council must rely on conflict and partisan mutual adjustment to attain information not otherwise obtainable. And the more complex the fishery, the higher the probability that the council will appear conflictful and the more such conflict will be perceived as counterproductive by those using a synoptic perspective* (e.g., the National Marine Fisheries Service).

[23] Lindblom *Politics and Markets*, p. 254.

Without the advantages of partisan mutual adjustment, nonbiological issues, which have been driven to the margins by time and resource constraints, would receive inadequate treatment, and the probability of approaching optimum yield would thereby be reduced. Thus, the process of partisan mutual adjustment, which is inherent in the multiplicity of decision points specified in the Fisheries Conservation and Management Act, plays the essential role of stimulating diverse preferences over a range of nonbiological issues.

Reliance on such processes can, however, cause delays in the formulation, approval, and implementation of the regional management plans. This development and approval process is complex and involves four basic stages: (1) presubmission, (2) discussion paper, (3) draft environmental impact statement, and (4) final environmental impact statement. To implement the fishery management plan after secretarial approval, the council or the National Marine Fisheries Service prepares proposed regulations.

While partisan mutual adjustment processes are at work in each of these phases they require a great deal of time and effort on the part of the councils and the Department of Commerce; this may contribute to delays. Since I expect the Department of Commerce to minimize its decision-making costs while attempting to get management plans in place, I hypothesize that *the greater the reliance on partisan mutual adjustment processes to ascertain preferences, the higher the probability of delay in the development and approval process. And the greater the delay, the higher the probability of conflict between the Regional Fishery Management councils and the Department of Commerce.*

Moreover, the differing responsibilities, structural dynamics and orientation of the Regional Fishery Management Councils vis-à-vis the National Marine Fisheries Service are basic sources of tension and, in some cases, result in outright conflict between the two. I suspect, therefore, that the professional-scientific training of most Service personnel leads them to adopt a synoptic view of management and decision-making. This orientation assumes clear technical criteria for judging the correctness of a management decision. Such decisions are seen as turning on questions of fact calling for scientific diagnosis. Conversely, the councils appear to approximate a partisan mutual adjustment process. The regional management councils were explicitly designed to substitute problem-solving interactions for conclusive diagnosis and analysis. The councils do not approach the problem looking for a "correct" solution but rather attempt to discover a position via competition among individuals with differing preferences in light of incomplete basic information. Their approach is basically a strategic one in terms of which they attempt to *satisfice*. This approach is highly "rational" owing to the requirement that the councils make decisions despite the fact that the management system as it is currently designed by the federal government does not have the formal capacity to generate the information required to calculate optimal yield. Yet the councils must do the best they can. This approach, by necessity, is not technical and has no correct solution in terms of which to judge the ultimate council decision. The fishery management process is fundamentally exploratory and remedial and explicitly accepts the limited capacity of the analyst.

The National Marine Fisheries Service searches for consistency in regulations across councils under the assumption that some degree of uniformity is required if the goal of optimal yield is to be reached. Such an approach gives the appearance of fairness or even-handedness, and has the added advantage that consistency or uniformity reduces transaction costs for the National Marine Fisheries Service.

The regional management councils will resist such an approach by arguing that each region has different types of fisheries and fishery industries, all of which have different characteristics and problems. They will argue that the drive for consistency will impose a potentially harmful uniformity on council decisions by downplaying such important differences.

Again, it is the organizational structure of the Regional Fishery Management Councils and the accompanying partisan mutual adjustment process that encourages the revelation of such differing preferences within and among regions. Yet if such fundamental differences were incorporated into the management plans, this has the potential of creating massive decision-making costs for the National Marine Fisheries Service, and the Department of Commerce, hence the drive for consistency.

Because of this synoptic approach to decision-making which has been hypothesized for the NMFS, I expect communication between the NMFS, the Department of Commerce, and the regional management councils to be strained. The NMFS will attempt, insofar as possible, to cast the discussions in purely technical terms. They will stress optimal yield as the objective to be reached and attempt to lead the councils to adopt a similar approach. This will be done despite the fact that the data to make such a determination with a degree of certainty is not currently available. While the NMFS will stress the necessity of reaching optimal yield, the councils will note that adequate data do not exist for economic, social, and ecological factors. Because of the paucity of such data the NMFS will argue that discussions should be largely confined to maximum sustainable biological yield, or at most, maximum sustainable yield adjusted for economic factors. The councils will, however, try to introduce a broad range of nontechnical issues which they feel are relevant to management plans and quota determinations.

The NMFS is expected to argue that while these nonbiological, nontechnical issues are undoubtedly important, they should not be given a weight equal to biological issues in the management plans. While the Service seeks to narrow the range of issues to purely technical matters, the management councils will try to broaden the range to nonbiological, nontechnical issues of concern to members and their constituencies.

It should be noted that there is a potential present for the National Marine Fishery Service and the Department of Commerce to carry off what might be called a "synoptic sleight of hand." This can be done by arguing that the management plans should be confined largely to technical issues as they relate to the overall goal of optimum yield. They will argue further that although optimum yield must be based on a large number of nonbiological variables, the focus will necessarily have to be narrowed because of an inadequate data base for most of the nonbiological factors, with the possible exception of economics. While the NMFS will be comprehensive and technical in their approach, pragmatism requires the exclusion of nonbiological factors; this skews the argument in a biological direction. In this sense, the synoptic approach potentially serves as a "razor" to cut away factors other than biology, which, in turn, tends to serve the predominantly biological interests of the professionals in the Service.

Ironically, the degree to which such a synoptic approach is utilized in the aforementioned fashion, lowers the probability that optimal yield will be approached. It is only through the "messy" process of partisan mutual adjustment that the diversity of preferences required for optimal yield can be ascertained and incorporated into management plans.

Despite the fundamental tension between these two approaches, a great deal of bargaining takes place among the Department of Commerce, the NMFS and the eight regional fisheries councils. The characteristics and properties of such bargaining and adjustment are, however, not well understood and remain a potentially fertile ground for research.

In this research effort different types of adjustment may prove to be the mechanism whereby some degree of coordination takes place among the strategic problem solvers in the regional management councils and the Service. We mean by coordination:

> A set of decisions is coordinated if adjustments have been made in it
> such that the adverse consequences of any one decision for other

decisions in the set are to a degree and in some frequency avoided, reduced, counterbalanced or outweighed.[24]

Lindblom also notes that when many decision-makers pursue problem solving as a strategy, they greatly strengthen the process of attending to neglected adverse consequences. In this sense, coordination can occur but only in a weak form.

The interaction and conflict among the Service and the Regional Fishery Management Councils may prove to be a valuable means of coordination insofar as each group is alert to values which they fear will be overlooked by others; each group is sensitive to consequences of decisions or policies that appear as marginal to others. In this sense, no one group can anticipate all of the positive or negative consequences of any decision.

Lindblom specifies three different types of bargaining and adjustment which may be present in this process: (1) *calculated adjustment*, (2) *deferential adjustment*, and (3) *parametric adjustment*. In calculated adjustment,

> X considers repercussions for Y before making his or her decision, even though he is able to make his decision without Y's cooperation. It is the kind of decision-making in which X's decision is designed to reduce or avoid injury to Y, or to offset injury with benefit so that Y's decision will not be disadvantageous to X.[25]

Deferential adjustment is defined as intended to avoid adverse consequences for another decision-maker. In parametric adjustment

> The decision-maker simply takes another's decisions as he finds them and adapts to them without deferring to or calculating repercussions on others. The adverse consequences of others for him are not forestalled, but they are reduced in effect—at an extreme wiped out— by his autonomous decision to make the best of the situation for himself. X reduces adverse consequences to himself by making a decision designed to cope with adversity to his own advantage.[26]

Which of these varieties applies to the relationships between the regional fisheries councils and the Department of Commerce, given variances in complexity, is an empirical question. But it is a distinct possibility that different forms of adjustment may be employed by the Secretary of Commerce in relation to different regional councils, depending on complexity.

[24] Lindblom, *The Intelligence of Democracy*, p. 154.
[25] Ibid., p. 158.
[26] Ibid.

Summary and Conclusions

My objective in this paper was to set forth an approach to understanding the complex process by which decisions are made by the regional fishery management councils. After delineating the objectives and structure of the Fishery Conservation and Management Act I noted the enormous analytic and informational requirements of the act. I then proceeded to argue that to understand how the structure and analytical requirements were interrelated in the decision process required an examination of two models of the decision-making process; namely, strategic and synoptic. In discussing the synoptic approach, I followed a line of reasoning akin to the following by Steinbruner:

> One cannot demand that decision-makers have perfect information as did the classical theory of the firm, for actual decision-makers never do. One cannot demand that they make impossibly elaborate calculations in making choices. The requirements which the assumptions impose must be to human dimension.[27]

But it is precisely such unreasonable requirements that the synoptic decision-making model makes. As a result, I argued that such an approach cannot be an adequate guide to understanding decision making under conditions of imperfect information, uncertainty, and risk. Yet these are the conditions faced by council decision-makers.

The strategic approach to decision-making, by contrast, uses assumptions which recognize man's limitations in a world of changing preferences and objectives. By drawing on a multiplicity of such strategic actors a policy which minimizes error can be forthcoming. This is possible by a process of interaction among these actors such that the adverse consequences of one group's decisions are called to the attention of the others. In this sense, a multiplicity of strategic actors interacting with one another copes with error in complex decision-making situations.

I used the strategic model to develop a series of testable hypotheses for a comparative study of the councils. It was suggested that such a study would be a first step toward the development of a positive model of regional fishery council decision-making.

The Fishery Conservation and Management Act can be characterized as an interesting mix of the synoptic and strategic approaches to decision-making. The stated objective of optimum yield is fundamentally synoptic in that it requires the finding of scientifically "correct" solutions to the fishery management problem. And it calls on the NMFS to harness their skills for this task. Nowhere is it noted that the attainment of such a goal

---

[27] Steinbruner, *Cybernetic Theory of Decision*, p. 27.

is impossibly costly in terms of time and staff requirements. In any case, the data required for such calculations do not currently exist for nonbiological factors. As a result, the comprehensiveness required by optimal yield cannot be achieved solely through scientific technical procedures.

It is a tribute to the designers of the Act, however, that they built in compensation mechanisms to these analytic requirements. The objective can be approached strategically by utilizing the organizational structure specified in the Act. This structure draws on the advantages derived from the interactions of a variety of actors which are worked out in terms of this overall management objective via a process of partisan mutual adjustment within the councils, between the councils, and between the councils and the National Marine Fishery Service. In this sense, the councils operate strategically with the working assumption that the analytical requirements of the act are very difficult, in short supply, and are prepared by individuals with varying levels of competence. For the councils, as compared with the NMFS, analysis is quite limited in its aspirations.

The process of partisan mutual adjustment which follows from the Act has the advantage that it *permits the analytic requirements of the Act to be adapted to problem-solving interactions.*

> Interactions are a substitute for analysis in the specific sense that they constitute processes that produce decisions in circumstances in which the decision cannot be or is not, for any reason, reached exclusively through analysis.[28]

and

> This process is used to stimulate what is otherwise an inadequate supply and quality of analysis.[29]

The organizational structure of the Fisheries Conservation and Management Act takes advantage of such processes to reach outcomes which they would not be competent to reach otherwise. Such interaction serves to keep incentives for problem solving widely dispersed and participation possibilities open. The latter also encourages the anticipation of negative consequences through the struggle between divergent views.

Ironically, however, it is the synoptic requirements of the Act which have received the greatest scholarly attention. Discussion constantly turns on the acquisition of more and improved data in terms of which to calculate optimal yield. While this is an important goal, this paper has

[28] Lindblom, *Politics and Markets*, p. 254.
[29] Ibid., p. 256.

directed attention to the merits of the strategic approach to decision-making which is at least as important as the scientific-technical planning components. The ways in which strategic problem solving substitutes for synoptic analysis is a subject worthy of extensive empirical research. This paper suggests a conceptual approach and some hypotheses in terms of which such a comparative analysis of the councils and their interactions with the federal government might proceed.

STUDYING OCEAN POLICY

CHAPTER 13
THE PROSPECTS FOR MARINE POLICY RESEARCH

LAURISTON R. KING*

Introduction

Rapid advances in ocean technology and an accelerating recognition that ocean issues are, in fact, a tangle of broad policy issues such as energy, pollution, and renewable resources have sparked a growing appreciation for the complexity of marine affairs. This awakening has not, however, been translated into sustained funding support for marine policy research. Nor are the prospects very good that support for marine policy research will increase much in the foreseeable future. This discussion suggests several conditions that have inhibited federal agencies from providing significant financial support for academic research in marine policy, reviews several variations of research and policy guidance, and concludes with some speculations about the probable fate of ocean policy research in the political arena.

Why a Bull Market for Marine Research is Not Likely

Marine policy research is the systematic analysis of the "goals, directives, and intentions formulated by authoritative persons and having some relation to the marine environment."[1] It is not, however, a field of study that is likely to grow dramatically. Despite modest increases in funding support and a number of scholarly publications during the 1970s, marine policy research has remained little known as a specialized field of study. The clientele for ocean policy studies is small and scattered, and agency managers and policy-makers are less than enthusiastic in their support for this kind of research.

One condition which will constrain the growth of support for marine policy research is the absence of a common identity among ocean users. Although shipping, fishing, dumping, and defense interests have the sea in common, these are all discrete activities with few powerful incentives for collective action. Few alliances exist between different sectors of the ocean community. Rarely do these users perceive their common political

---

* Lauriston R. King is Deputy Director of the Sea Grant Program and Assistant Professor of Management at Texas A&M University.
[1] John King Gamble, Jr., *Marine Policy: A Comparative Approach* (Lexington, Mass.: Lexington Books, 1977), p. 6.

interests in the formation of national ocean policy. Most do not even provide for national representation through Washington-based trade and professional associations as do most established and emerging interest groups.

The effect of this lack of cohesion on marine policy research is to scramble the focus on ocean affairs. Potential researchers have found it hard to identify those aspects of a broad issue which deal with the ocean. Sorting out and understanding these ocean issues requires time and a diversion in career interests—big investments for which prospective policy researchers are given little incentive.

The segmented character of the ocean community is mirrored in the federal organization for marine affairs. In 1975, for example, federal marine science and marine-related activities were carried out by 11 organizations in six departments and five agencies.[2] Amidst this proliferation there exist only a few potential patrons for marine policy research: the National Oceanic and Atmospheric Administration (NOAA), the National Science Foundation (NSF), the United States Geological Survey (USGS), the Environmental Protection Agency (EPA), and the United States Navy (USN).

None of these agencies, however, has committed itself to long-term support of marine policy research. In fact, federal support for marine-related social sciences—the key element in most marine policy studies—has been paltry at best. Between 1970 and 1975, federal agencies spent nearly $11 million on 304 projects in the marine-related social sciences—about 0.3 percent of the $3.7 billion ocean budget for that period. When compared to the "research" category of the federal ocean program, support for marine-related social sciences accounted for only about 1.7 percent of the $648.1 million total. Almost all of the remaining funds went to support research in oceanography and ocean engineering.[3] An examination of the kinds of social science research projects that are supported is also revealing. Most of the projects were applied research aimed at resolving a fairly specific short-term problem. They drew on a variety of disciplines including engineering and natural sciences to analyze topics such as deepwater ports, aquaculture, or coastal zone management.[4]

[2] United States Comptroller General, *The Need for a National Ocean Program and Plan*. Report to the Congress, October 10, 1975. GGD-75-97, p. 1.

[3] *Federal Agency Support for Marine-Related Social Science Research*. Report from an *ad hoc* subcommittee for the Interagency Committee on Marine Science and Engineering, December 1976, p.2. Social science can function as something of a surrogate for policy studies, since most of these studies are either conducted by or draw heavily upon skills from economics, political science, sociology, and anthropology. Note, however, that not all social science research described in this report can count as policy research.

[4] *Ibid.*, p. 3.

A recent study of support for marine-related social science research by NOAA's Sea Grant Program revealed that between 1974 and 1978 (the years for which comparable data were available), support for social science research was a modest part of Sea Grant's overall budget. The approximately $2.0 million that went to the social sciences amounted to no more than 15 percent of Sea Grant's shrinking research budget for any given year. The amount of money allocated for research dropped from more than 70 percent to slightly over 50 percent of the total Sea Grant budget during these years. Typically, the yearly percentage for social science research was even smaller, fluctuating between 10 and 13 percent.[5]

Aside from budgetary constraints, the predominance (until very recently) of scientists and technicians in the ocean agencies has inhibited active support for policy work. The professional backgrounds of an agency's staff contribute significantly to the development of an agency culture—an "internal set of loyalties and values which are likely to guide its actions and policies."[6] Bureaucratic culture plays a critical role in shaping the way an agency sees its job, the way it views the legitimate scope of its concerns, and the way in which the skills and people required to carry out its mission are selected. For the ocean-related agencies, professional staffs have tended to see their responsibilities primarily as the collection, generation, and dissemination of scientific data and analysis, not the resolution of highly charged issues of the use and management of ocean resources.

NOAA, for example, as a technically oriented mission agency, has been composed of middle and upper management personnel with training in the oceanographic, atmospheric, and related sciences. In some respects NOAA remains a loose confederation of groups involved in environmental research and services, with little experience or familiarity with policy studies. This lack of familiarity with policy studies or experience in using them, together with an agency culture that is biased toward the technical aspects of the agency's mission, make the prospects for new initiatives in support of marine policy research most unlikely.

The influence of a scientific agency's culture was seen in efforts to set up a marine science affairs program in the National Science Foundation's Office for the International Decade of Ocean Exploration (IDOE). The IDOE, the major international oceanographic research program in the United States during the 1970s, was designed to support large-scale, long-

---

[5] Michael K. Orbach and Lauriston R. King, *The Social Sciences in the Sea Grant Program*. A report prepared for the Research Committee of the Sea Grant Association, October 24, 1979, table 1, p. 6.

[6] Harold Seidman, *Political Power, and Position—The Dynamics of Federal Organization*, 2nd ed. (New York: Oxford University Press, 1977), p. 66.

term multidisciplinary projects in environmental forecasting, seabed assessment, environmental quality, and living resources.[7] The rationale for an IDOE marine science affairs program was that the scale and effort of these projects promised to accelerate understanding of the marine environment and its resources at a rate which would challenge the ability of the social and political system to cope. Its goal was to build into the scientific research program a way to translate and interpret scientific findings into terms understandable to those outside the immediate ocean research community.[8]

The NSF professional staff was restrained in its enthusiasm. This was not surprising since marine policy research as a specialized field of study had only just begun to take hold in the early 1970s. There was no cadre of specialists to press for sustained research support, nor was there an established and visible body of scholarly literature that showed what kind of work would result from a marine science affairs program. Moreover, the kind of science and technology assessments envisaged for the program had few, if any, precedents in the history of oceanographic research. There were simply no examples to help skeptical oceanographers and NSF bureaucrats understand the proposed marine science affairs research program.

These difficulties were compounded by the presence of an agency culture which remained passionately committed to the support of basic research in the natural sciences. The notion of a coherent program of policy research was alien to many members of the NSF staff. For example, in the 10 years between 1961 and 1971 the NSF budget for marine science increased from $4.7 million to $41.8 million. Despite this dramatic increase, NSF essentially disregarded the social, economic, legal, or political implications of new discoveries and technologies in ocean science. More than 98 percent of NSF funds through 1972 went to traditional oceanographic research projects, special programs, and vessels

---

[7] On the organization context for the International Decade of Ocean Exploration, see Peter F. Hooper, "The Administrative Policy Process for Science: A Case Study of Organizational-Environmental Dynamics" (Doctoral Dissertation, University of Connecticut, 1979). For a retrospective look at the IDOE, see *Oceanus* 23 (Spring 1980):*passim.*

[8] The rationale and goals of the program are described by Lauriston R. King, "The International Decade of Ocean Exploration Program in Marine Science Affairs," *Marine Technology Society Journal* 11(1977):10-15. The difficulties of starting the Marine Science Affairs Program were further compounded by the prominent role played by the social sciences in policy studies. NSF has historically had a hard time with the social sciences. Public officials and tax crusaders have frequently singled out NSF social science projects as targets for ridicule. The prospect of tying the social sciences approach to potentially visible and controversial policy studies did not provide a powerful incentive to press for a program of marine policy research, particularly in a field as small, fragmented, and ill-defined as marine affairs.

and facilities. Only one-half of one percent was directed to policy and planning studies about the marine environment. The proportions did not change significantly during the late 1970s.

The organization's bias toward natural and physical science was also evident in the expression by oceanographers of their concern about the prospect of diverting IDOE funds from basic scientific research projects to marine science affairs. Several oceanographers argued that the proper business of the IDOE was to support the traditional marine sciences. The undisguised sentiment was that support for marine policy studies was not the best use of funds and that these funds would be better spent on basic oceanographic research. In short, it was hard to market an idea for which there was no proven demonstration model, no demand, and no solid base of institutional support.

Increased marine policy research has not received much encouragement from top agency managers either. Despite the judgment of many ocean policy specialists that marine-related issues will grow in scope and complexity, the oceans do not occupy an especially prominent position on the agendas of political elites. There are simply not enough incentives for elites to invest their time or resources on ocean issues except in times of crisis or under intense political pressure. In order to reduce their investments in time, money, political prestige, and good will, prudent decision-makers will try to cope with short-term, (ostensibly) manageable issues, rather than more vaguely defined long-term issues.[9]

What Then Are the Prospects for the Use of Marine Policy Research?

Marine policy studies may play a role in influencing policy-makers, but it is more likely that these studies will meet a set of political needs for policy-makers other than rational enlightenment. Studies will buy time, give the appearance of action, help them keep their options open, and, if timing is right, reinforce or legitimize actions the decision-makers might take without the benefit of learned academic studies. By sorting out different "families" of policy research, it is at least possible to speculate about the prospects for use. These "families" include advisory reports, short-term solicited projects, formal policy research and general historical and analytical studies.

[9] Davis B. Bobrow, "International Politics and High-Level Decision-Making: Context for Ocean Policy," *Ocean Development and International Law* 3(1975):171-180.

Advisory Reports

Historically, this approach to policy guidance has been a favorite in the oceans arena. Typically it involves an almost exclusive reliance on individual experience and knowledge exchanged at workshops, seminars, advisory meetings, or symposia. Recommendations are based on negotiation, not on classically documented empirical studies. Prominent examples include the reports of the National Advisory Committee on Oceans and Atmosphere, the National Academy of Science's Ocean Policy Committee and Ocean Sciences Board, and the National Academy of Engineering's Marine Board.[10]

Using advisory committees for policy guidance has served a variety of purposes in the political marketplace. It has been one way to get expert advice at low cost. Advisory groups usually come cheap—plane tickets, meals, and lodging—and can address issues of specific concern to agency sponsors and produce recommendations in short order. In doing so they can contribute to building a historical record of policy recommendations that can be invoked as the need or opportunity arises. Emphasis is on results in the form of recommendations, not on the analytical base or the academic craftsmanship with which such cases are put together.

The prospects that this approach to policy guidance will persist are very good, mainly for the same reasons it has prospered to date. It calls for small investments of time and money, brings together experts on a particular issue with little commitment to their long-term care and feeding, and builds a reservoir of recommendations available to elites as they need them.

---

[10] Examples include: *International Marine Science Affairs: A Report by the International Marine Science Affairs Panel of the Committee on Oceanography*, National Academy of Sciences, Washington, D.C.; *OCS Oil and Gas: An Assessment of the Department of the Interior Environmental Studies Program*, a report to the Department of the Interior, Bureau of Land Management, from the Committee to Evaluate Outer Continental Shelf Environmental Studies, National Academy of Sciences, Washington, D.C., 1978; *The Continuing Quest: Large Scale Ocean Science for the Future*, prepared by the Post-IDOE Planning Steering Committee, Ocean Sciences Board, National Academy of Sciences, Washington, D.C., 1979; *Port Development in the United States*, a report prepared by the Panel on Future Port Requirements of the United States, National Academy of Sciences, Washington, D.C., 1976; *The Agnes Flood: a Post-audit of the Effectiveness of the Storm and Flood Warning System of the National Oceanic and Atmospheric Administration*, a report for the administrator of NOAA by the National Advisory Committee on Oceans and Atmosphere, November 22, 1972; *Reorganizing the Federal Effort in Oceanic and Atmospheric Affairs*, 2 vols., a special report to the President and the Congress from the National Advisory Committee on Oceans and Atmosphere, February 1979; *Perspectives on Ocean Policy*, Conference on Conflict and Order in Ocean Relations, October 21-24, 1974, Airlie, Virginia; Lee G. Anderson, ed., *Economic Impacts of Extended Fisheries Jurisdiction, Proceedings of a Symposium Sponsored by the University of Delaware Sea Grant Program and the National Marine Fisheries Service*, 1977.

Short-Term Solicited Projects

Another category of ocean policy research is the short-term, solicited problem solving often associated with agency contracts and congressional requests. These kinds of studies usually are well-defined in advance, the boundaries of work are described in Requests for Proposals (RFPs), and the outcome is anticipated in advance. Typically, these short-term studies are produced quickly and are paid for by the agency sponsor. Researchers have played only a small part in defining the problems to be addressed, and hence assume the role of tradesmen working under contract to solve a problem or provide information on subjects that may be of incidental interest to them.

Marine policy research of this sort has been conducted at the request of Congress by the Congressional Reference Service, the Office of Technology Assessment, and the General Accounting Office, as well as by the major ocean agencies.[11] The studies tend to be descriptive rather than conceptual, derivative rather than original, and often transient in their use. A premium is placed on clarity of expression, timely completion, and immediate availability for policy deliberations. Rarely do these kinds of studies command the respect that academic policy researchers accord to their peers who are engaged in ostensibly more rigorous, systematic variants of the genre.

Increased pressure on the ocean agencies for action and decisions on complex issues of use and management will continue to sustain the demand for problem-solving studies. Issues can be reasonably well defined. Projects can be funded on an *ad hoc* basis without any long-term funding commitments by the mission agencies. And, there is a large reserve of knowledge-producers in the universities, think tanks, and consulting firms eager to respond to agency solicitations.

[11] Examples include: *Export and Domestic Market Opportunities for Underutilized Fish and Shellfish*, prepared for the U.S. Department of Commerce by Earl R. Combs, Inc., December 1978; *Coastal Effects of Offshore Energy Systems—An Assessment of Oil and Gas Systems, Deepwater Ports, and Nuclear Powerplants off the Coast of New Jersey and Delaware*, United States Congress, Office of Technology Assessment (November 1976); *North Sea Oil and Gas: Impact of Development on the Coastal Zone* (October 1974), and *The Economic Value of Ocean Resources to the United States* (December 1974), both prepared at the request of Senator Warren G. Magnuson, chairman, for the use of the Committee on Commerce, pursuant to Senate Resolution 222, National Ocean Policy Study (Washington, D.C.: U.S. Government Printing Office, 1974); *Soviet Ocean Development* (Washington, D.C.: U.S. Government Printing Office, 1976), prepared at the request of Senator Warren G. Magnuson, chairman, Committee on Commerce, and Senator Ernest F. Hollings, chairman, National Ocean Policy Study, pursuant to Senate Resolution 222, October 1976.

Systematic Marine Policy Studies

Compared to short-term solicited studies, this class of work usually is initiated by investigators (often through negotiations with potential consumers) and tends to be more intellectually rigorous, thorough, and deliberate in its weighing of alternatives. It also tends to be less solicitous of policy-makers' demands for rapid response and explicit guidance on specific policy issues.[12] It is the kind of study that emerges from the premise that rigorous, critical, and systematic analysis can provide the basis for more "rational," and hence better ocean policy.

The key assumption is that in trying to manage and understand big environments like the oceans and the atmosphere, things do not always work out the way they are supposed to. As comic strip character Willie Mullins is made to say, "Results is what you expects and consequences is what you gits." Careful problem definition, data collection, and scientific analysis is presumed to be a far better basis for making public policy than reliance on speculation, guesswork, and gut reaction. The presumption is that man's ties to and use of the oceans are likely to be positive and constructive only to the extent that policies are guided by reliable knowledge about the marine environment. In short, these are the kinds of disinterested studies that are promoted by academic students of marine affairs and by political scientists in particular. They also are the kinds of studies that frustrate politicians because of their conceptual trappings, equivocal recommendations, and apparent lack of availability when they are required.

Because the more formal, systematic approach to policy analysis is preferred by academics, the burden for selling this approach usually falls

---

[12] The remaining categories, systematic policy studies and historical and analytical works, obviously are far from distinct, and therefore some illustrations may seem a bit arbitrary. Nonetheless, it is possible to at least suggest some examples of work in the respective research families. Among the more systematic, self-conscious policy studies have been: Don E. Kash, et al., *Energy Under the Oceans—A Technology Assessment of Outer Continental Shelf Oil and Gas Operations* (Norman, Oklahoma: University of Oklahoma Press, 1973); *Effects on Commercial Fishing of Petroleum Development Off the Northeastern United States*, a report from the Marine Policy and Ocean Management Program, Woods Hole Oceanographic Institution, Woods Hole, Mass. (WHOI-76-66, April 1976); R.E. Osgood, A.L. Hollick, C.S. Pearson, J.C. Orr, *Toward a National Ocean Policy: 1976 and Beyond*, Ocean Policy Project, School of Advanced International Studies, Johns Hopkins University (Washington, D.C.: U.S. Government Printing Office, 1975); M. Hershman, R. Goodwin, A. Ruitsala, M. McCrea, Y. Hayuth, *Under New Management—Port Growth and Emerging Coastal Management Problems*, a Washington Sea Grant Publication distributed by University of Washington Press, Seattle and London, 1978; S. Brown, N.W. Cornell, L.L. Fabian, E.B. Weiss, *Regimes for the Ocean, Outer Space, and Weather*, (Washington, D.C.: The Brookings Institution, 1977).

on them. They must persuade often diffident agency managers that an investment in a project proposed from outside the agency will redound to their benefit. But because academics do not always take on issues that appear to have immediate bearing on a problem, these studies will continue to be less attractive investments for the ocean agencies than research on shorter-term issues requiring prompt resolution.

Historical and Analytical Case Studies

Advisory reports, short-term solicited studies, and systematic analyses respond to more or less clearly perceived issues and well-defined (or at least discernible) groups of potential consumers, and offer recommendations geared specifically to the needs of policy-makers.

Historical and analytical case studies are not produced to serve the needs of the powerful but rather they reflect the personal interests of the researcher. Policy recommendations, where they are offered, are presented in general terms in the context of the historical evolution of the particular issue.[13] This approach to marine policy research draws on formally documented reconstructions of political settings and policy decisions. It is often descriptive rather than analytical and aims less at explicit policy advice than at understanding the political context in which the advice must be considered. It seeks to build the intellectual foundations for marine studies by structuring and recreating the political, institutional and legal conditions that brought us where we are.

The main clientele for these kinds of studies is not the policy-makers now in positions of bureaucratic or political power, but students who are likely to assume these positions. But because this clientele is diffuse, at the periphery of power, and not commonly seen as a consumer for policy study and guidance, it is not in a position to urge increased support for this kind of research. There are few incentives for sustained support of

[13] A sample would include: Robert L. Friedheim, *Understanding the Debate on Ocean Resources*, University of Denver Social Science Foundation and Graduate School of International Studies monograph series in world affairs, 1969; S.A. Lawrence, *International Sea Transport: The Years Ahead* (Lexington, Mass.: Lexington Books, 1972); Edward Wenk, Jr., *The Politics of the Oceans* (Seattle and London: University of Washington Press, 1972); Gerard J. Mangone, *Marine Policy for America* (Lexington, Mass.: Lexington Books, 1977); Ross D. Eckert, *The Enclosure of Ocean Resources* (Stanford, Ca.: Hoover Institution Press, 1979); E.R. Pariser, M.B. Wallerstein, C.J. Corkery, and N.L. Brown, *Fish Protein Concentrate: Panacea for Protein Malnutrition?* (Cambridge, Mass. and London: MIT Press, 1978); Harvey B. Silverstein, *Superships and Nation-States: The Transnational Policies of the Intergovernmental Maritime Consultative Organization* (Boulder, Colo.: Westview Press, 1978); Ann B. Hollick, *U.S. Foreign Policy and the Law of the Sea* (Princeton, N.J.: Princeton University Press, forthcoming).

historical and analytical studies. This is especially true when those in positions to support marine policy research perceive their most pressing needs to be short-term studies directed toward well-defined consumers, with some prospect for influencing policy.

Despite the differences between families of marine policy advice, the underlying assumption for all of them is that somehow the advice will be used. In fact, one persistent delusion held by proponents of policy analysis is that policy research *per se* is sufficient to shape prudent, rational public policy. It is not. The quality, epistemic craftsmanship, and analytical rigor of academic research translates only roughly into the social and political context of policy debate.[14]

Knowledge—in the form of policy analysis—is rarely its own master. This is nowhere more true than when it becomes embroiled in political conflict. Knowledge does not come to politics in pristine form. It is shaped, manipulated, and often distorted by the perceptions of interested policy-makers, the political demands of the policy setting, and the way the knowledge was derived.[15] Accidents of timing, political interest, public visibility, and individual personalities often have more to do with the influence of policy studies on policy-makers than does academic rigor. Two recent studies, one on development of the outer continental shelf (OCS), and the other on predicting outcomes at the Law of the Sea Conference, illustrate the point.

The continental shelf study was carried out by a multidisciplinary group from the University of Oklahoma Science and Public Policy Program and was funded by the National Science Foundation in 1971. The team deliberately selected outer continental shelf (OCS) oil and gas development for its anticipated policy importance during the period in which they expected to conduct their research.[16]

By most measures the project was a model of the ideal link between policy research and political decision-makers due in part to the credibility of the researchers and their substantive understanding of OCS technology and policy issues, to continued support by NSF after the final report was

[14] "Epistemic craftsmanship" is a term used by political scientist Philip H. Melanson as a "shorthand designation for an approach to attaining valid explanation and reliable knowledge." He suggests that "unless logical models or processes of explanation are geared to the phenomena, goals, and scientific exigencies of an area of inquiry, they are likely to be of little use and to be little used. Only when they are contextually meaningful can they do what they are supposed to do: enhance the validity of knowledge." See Melanson, *Political Science and Political Knowledge* (Washington, D.C.: Public Affairs Press, 1975), pp. 60-62.

[15] Lauriston R. King and Philip H. Melanson, "Knowledge and Politics: Some Experiences from the 1960s," *Public Policy* 20(Winter 1972):83-101.

[16] Don E. Kash, "A Report on Technology Assessment of OCS Oil and Gas Operation," paper presented at the American Political Science Association, Chicago, September 1974.

distributed, and, in large part to historical chance. To allay the doubts of both environmentalists and industrialists that they were advocates for one side or the other, the group judiciously appointed an oversight committee that included representatives from Congress, the Executive Branch, the petroleum industry, and environmental interest groups. They also dealt with consultants from every segment of the policy network; conducted recurrent interviews with experts from the OCS oil and gas community; distributed draft papers widely; and conducted a one-week workshop involving 80 representatives from all parts of the OCS community. These activities nurtured a clientele for the final report, *Energy Under the Oceans*,[17] and established the credibility of the research team.

The impact of the well-reviewed study on public policy was, however, due largely to luck and chance. Two days after the report was formally delivered to the National Science Foundation, President Richard Nixon proposed that OCS leasing be tripled and that the Council on Environmental Quality conduct a one-year study of OCS oil and gas operations. *Energy Under the Oceans* became one of three major background documents for a series of six public hearings which were scheduled around the country. The visibility of the report was further enhanced when NSF purchased 1,500 copies for free distribution and extended financial support to ensure the team's availability for consultation. The Arab oil embargo and the rapid increase in attention to offshore oil and gas gave the final impetus to widespread attention to the study.

As an effort to influence policy, the OCS study was a success for a variety of reasons, not all of which were due to the quality of the report. First, the Oklahoma team was seen as being knowledgeable about OCS oil and gas operations but without a stake in the outcome. Second, because they were perceived as disinterested, they collaborated effectively with different members of the ocean community. Third, chance and timing played a critical part in drawing attention to the report. The Arab oil embargo, the presidential initiative on the OCS, and the CEQ study could hardly have been anticipated at the start of the project. Fourth, with support from NSF, the members were available to discuss their work with all interested groups for some time after the report was released—a rare luxury for most policy studies. Finally, the study made 39 specific recommendations that could be (and were) designed for immediate adoption by both Congress and the executive agencies.[18]

The apparent success of the OCS project in terms of visibility and policy influence stands in sharp contrast to the experience of the Law of

---

[17] Kash, et al., *Energy Under the Oceans.*

[18] Kash, "A Report on Technology Assessment of OCS Oil and Gas Operations," pp. 9-19.

the Sea Forecasting Project. The project started in 1972 with a definition of a research task between the U.S. Navy and the Center for Naval Analysis, a Washington-based think tank. The goal of the project was to predict the broad directions of national bargaining in the United Nations Law of the Sea Conference using content analysis processed by sophisticated statistical and quantitative techniques. Although formal work began in 1972, interest in developing analytical tools for coping with large-scale multilateral negotiations had been expressed as early as 1969.[19]

Navy officials endorsed the CNA proposal, convinced of the potential value of efforts to order and process information from the deliberations, as well as to outline strategies, define alternatives, and weigh tradeoffs. When State Department staff learned about the project, they also contributed a modest part of the financial support. Although the amount was small, it justified direct contact between the State Department and the project investigators. Despite the investigators' experience, the sophistication of the research tools, and the existence of prospective users for the work, the forecasting project was limited in its influence by forces which had little to do with the skills of the researchers or the quality of their work. These included strained interagency relations, personnel changes, and communication difficulties with prospective users of project findings.

Since they were sponsored by the Navy and housed in the Center for Naval Analysis, it was inevitable that the project members would have difficulty in establishing their credibility as disinterested analysts. These difficulties were compounded by later developments in project reporting requirements. For example, the State Department failed to renew its share of the contract in 1974, and thus the only way their work could be officially disseminated was through the Navy to the Department of Defense hierarchy, and from them, at their discretion, to other members of the Task Force.[20] Not surprisingly, Navy officials adopted a somewhat proprietary attitude toward project information—an attitude which in appearance, if not in fact, inhibited the free exchange of data.

Command changes in top Navy ranks also caused an erosion of the close personal relationships between the researchers and naval officers responsible for protecting the Navy's interests in the negotiations. The uncertainties aggravated by personnel changes were reflected most clearly when a major user of project results, a senior member of the Defense Department's Law of the Sea Task Force, left the department.

---

[19] Robert L. Friedheim, "Research Utilization Problems in Forecasting the UN Law of the Sea Conference; or, The Perils of the Persian Messenger," in *Formulating Marine Policy: Limitations to Rational Decision-Making*, Proceedings, second annual conference, Center for Ocean Management Studies, University of Rhode Island, June 19-21, 1978, pp. 144-167.

[20] *Ibid.*, p. 150.

Organization—or its absence—also reduced the likelihood that well-grounded policy advice would be especially influential. The United States delegation to the Law of the Sea Conference was too cumbersome and too poorly organized to effectively use analyses from the forecasting project. The desire to protect agency interests was a powerful force for the selection and use of political intelligence. Moreover, there were not enough people with the technical skills and training needed to comprehend readily the information from the project. Nor had sufficient standard procedures for receiving, processing, and disseminating the information been established.[21]

Problems of organization were compounded by difficulties in communicating across disciplines. Different experience, education, values, roles, and time constraints shaped perspectives on what information was valuable. The contrast was particularly evident between lawyers and policy analysts. The lawyers put great stock in their personal negotiating skills and their knowledge of delegates and country positions. For them, experience, not research, was the key to the right approach for dealing with their adversaries and allies. Although the analysts respected experience and intuition in the bargaining process, their orientation as researchers was systematic and quantitative. "We believed in probability, and thought bargaining decisions should be strongly influenced, but not determined, by the odds. Such different mental sets obviously led to some hostility and misunderstanding."[22]

It is tempting to view the OCS and forecasting projects as examples of different degrees of success in influencing the policy process. That is not the intent. Rather, the two cases were offered as examples of policy research on visible and important public problems carried out by experienced, knowledgeable specialists. In neither instance, however, was the degree of intellectual rigor or scientific craftsmanship the key to influencing policy. Other factors over which the researchers had little control appear to have been far more important. These included timing, credibility of the researchers and their work, personal relationships, organizational structure, and sustained support from the sponsoring agency. The point is that while quality, sophistication, and adherence to the canons of scientific inquiry all are desirable in policy research, they are not guarantees that such research will influence policy in any predictable fashion.

---

[21] *Ibid.*, p. 161. For a more general view of the organizational problems confronting the United States in dealing with the Law of the Sea Conference, see Otho E. Eskin, *Law of the Sea and the Management of Multilateral Diplomacy*, Oceans Policy Study 1:15, Center for Oceans Law and Policy, University of Virginia, May 1978.

[22] Friedheim, "Research Utilization Problems," p. 165.

Concluding Observations

Despite the dismal prospects for any significant increase in funding for marine policy research, changes in the marine area continue which bode well for a modest level of support during the 1980s. These include burgeoning management and policy responsibilities imposed on the ocean agencies, marine affairs specialists moving into agency management positions, and a small but growing community of ocean scholars.

NOAA, for example, has assumed increasing responsibilities for resource management in addition to its more traditional environmental activities. Within its brief 10-year history, NOAA has been assigned responsibility for implementing the Coastal Zone Management Act (1972) (and more recently, expanded authority under the 1976 amendments), the Deepwater Ports Act (1972), ocean dumping under the Marine Protection, Research and Sanctuaries Act (1972), the Marine Mammal Protection Act, the Fishery Conservation and Management Act (1976), and the National Ocean Pollution Research and Development and Monitoring and Planning Act (1978). All these jobs, with the possible exception of ocean dumping, are regulatory or developmental tasks rather than scientific or technical activities. Other agencies such as the Geological Survey and the Environmental Protection Agency have been assigned expanded regulatory and management roles for specific marine activities as well.

The marketplace for policy guidance and advice will continue to expand as demands for agency performance in these areas continue to grow. Agency staffs will include specialists trained in marine policy and ocean management who are likely to be more receptive to policy research than are their counterparts in the natural and physical sciences. The academic literature in marine affairs will continue to grow, and a network of scholars with marine interests will continue to coalesce.

The best prospects for marine policy research are the advisory activities and short-term solicited research projects that meet the needs of those in power. These meet agency needs for fast response, frame a problem in terms defined by agency consumers, and do not involve long-term commitments after the work is completed.

Regrettably, the promise of slow but steady growth in these families of marine policy research is far more likely to have a corrosive rather than a constructive influence on the quality of discourse about marine policy. Because they address *ad hoc*, short-term issues, there is little time for original research or extensive data collection. Consequently, most of these studies must be highly derivative. They add little to the conceptual and empirical foundations upon which analysis and evaluation must be built. More important, *ad hoc* funding of short-term projects ignores any responsibility for nurturing a community of marine affairs specialists capable of anticipating and assessing problems over the long term.[23]

The greatest need, and the one least likely to be encouraged, is for long-term investments in the critical intellectual work of the kind found in systematic policy analyses and historical and analytical studies. Academic researchers are far less encumbered by preconceived definitions of a problem or policy issue. They are in a much better position to expose faulty assumptions, challenge conventional wisdom, criticize political decision-makers, and see beyond current issues to those lurking at the fringes of the existing policy agenda. Thoughtful, thorough public discourse on marine policy requires a clear understanding of the political, institutional, and legal conditions that form the contexts for current issues. As the range of marine policy issues expands and the oceans become even more intimately linked to broad issues like national security and energy, the importance of historical perspective and informed criticism in the formulation of sound policy must increase.

[23] Two ocean specialists have accurately assessed the dilemma of this short-term perspective:

> The consequence of this thinking leads to an emphasis on short-term research needs and places middle and long-term needs in the background. Here lies the dilemma. Activities are moving so rapidly in ocean-related areas that it is essential to keep abreast of the latest information. Yet this leaves little time for long-range thinking, and both current and potential activities could have such far-ranging implications globally as well as nationally, that we are forced to think both short-term and long-term. But how do we do this when the vehicles and institutions for implementing even short-term policy research projects are inadequate both from the funding agency and the receiver end of the system?

Robert L. Friedheim and Judith T. Kildow, *Report of the Ocean Policy Research Workshop*, National Science Foundation, Washington, D.C., February 11-12, 1975, p. 19.

CHAPTER 14
PROVIDING DIRECTION TO THE
NATIONAL OCEAN POLICY RESEARCH EFFORT

ROBERT L. FRIEDHEIM*

What is national ocean policy? Is a focus on national ocean policy a legitimate field of study? Are there knowledge gaps that should be filled, research trends that should be encouraged, or analytic methods that should be applied? Questions of whether national ocean policy exists as a distinct entity, where it exists, its comparative importance vis-a-vis other policy problems, its worthiness of being studied and having resources expended upon its study, and where these resources should be expended have occupied the thoughts of a group of ocean policy specialists for almost a decade and a half.

There has been a widely shared presumption, at least in the United States, that a national ocean policy does exist but that it lacks cohesiveness. It is also assumed frequently that the way to attain a cohesive and coherent ocean policy is to reorganize the United States government agencies that are responsible for ocean policy. Those in and out of the federal government who share such a presumption and who perceive a trend in world ocean policies toward ocean enclosure find it necessary, as a result, to ask two further questions: Is national ocean policy study a field, a subsector, or, by any other designation, a coherent subject of inquiry? Do we know enough about national ocean policy in the United States or in other developed states or even in developing states to not merely describe the trends in national ocean policies but also to prescribe wiser future courses of action.[1]

It has long been considered desirable by many to have an evaluation and assessment workshop focusing on the subset of ocean policy studies produced in academia. At the request of the Marine Science Affairs Program of the International Decade of Ocean Exploration of the National Science Foundation[2] I had the opportunity to convene such a

* Robert L. Friedheim is Associate Director for Marine Policy at the Institute for Marine and Coastal Studies, University of Southern California, and **Professor** of International Relations, University of Southern California. Support for the preparation of this chapter was provided by the Marine Science Affairs Program, International Decade of Ocean Exploration, National Science Foundation, Grant No. OCE 77-23092

[1] The concern with national ocean policy should be normative and prescriptive as well as theoretical and descriptive.

program. Twenty-six invited participants met from April 13 to April 16, 1978, at the Catalina Marine Science Center of the Institute for Marine and Coastal Studies of the University of Southern California. In this chapter, I will attempt to contribute to a refinement of the study of national ocean policy by reporting on the discussions at that meeting and the recommendations resulting from it.

The Mandate of the Workshop

The workshop was convened for the specific purpose of examining the state of the art in national ocean policy studies being done in academic institutions and making a recommendation as to whether the Marine Science Affairs Program should invite proposals on national ocean policy problems.

A number of scholars were assembled in order to ask whether our fundamental base of ocean policy knowledge is adequate to provide the necessary intellectual substructure for wise decisions by the next generation of governmental ocean policy decision-makers. It was clearly not practical to assemble a group of knowledge producers, mostly from universities, and knowledge consumers, mostly from governmental research management offices, to produce recommendations on what current research may pay off immediately in government policy. But if the oceans are to have a future, we must know more about the evolving nature of the system which provides the context for active decision; the patterns of behavior manifested by various interest groups, political parties, governments, and international organizations in making ocean policy; and the policy options available in the light of both human desires to mold the situation and the constraints imposed by the physical nature of the ocean. In other words, it would seem that basic ocean-related social science research is as necessary as basic hard science research if we are to understand, and perhaps master, the ocean world.

It was to achieve this end of developing basic ocean social science knowledge that the workshop was organized. The participants were chosen to be as representative as possible of producers of ocean knowledge (university-based ocean policy researchers) and consumers of ocean knowledge (U. S. government research managers). Not overlooked was the need for a range of knowledge among the participants concerning

---

[2] Dr. Lauriston King was the program manager at the time of the request and of the meeting. He subsequently left the National Science Foundation and has become the Deputy Director of the Sea Grant Program, Texas A&M University. Dr. Louis Brown has assumed Dr. King's responsibilities for the Marine Science Affairs Program.

the national ocean policies of the United States, other developed states in the European community, and Japan, the Soviet Union, and Eastern Europe, and the developing states. In terms of disciplines, ocean scientists as well as ocean social scientists were present since the Marine Science Affairs program was concerned with questions of the impact of ocean knowledge on public policy. Finally, to help avoid a parochialism that could result when "ocean people" talk with other "ocean people," social science theorists and analysts who have dealt with parallel but not ocean-related problems were invited to participate in the workshop.

The State of the Art in Analyzing National Ocean Policy

What is national ocean policy? An acceptable, formal academic definition is: "a set of goals, directives, and intentions formulated by authoritative persons and having some relationship to the marine environment."[3] It would include all activities relating to the substance of the uses of the oceans by nation-states: how they make their decisions, and how they organize themselves to make their decisions. [4] However, many critics ask, is there anything in the nature of salt water which makes it sensible to attempt to treat policy relating to the ocean as a separate and distinct entity? This question is often followed by the argument that there is no coherent policy for all land-related policy matters. While I would not claim that salt water *per se* makes ocean policy unique, I do believe that there is a distinctive physical attribute to salt water that causes policy related to it to have some important shared characteristics. The ocean (salt water) is a physical common. It is difficult to divide it up. It is difficult to prevent the currents from moving where nature intended them to move, or to stop fish from migrating to wherever they desire. Not long ago we used all of the assets of the ocean beyond the territorial sea in common. Recently, there has been a move toward enclosure by extending national jurisdiction outward from the

[3] John K. Gamble, Jr., *Marine Policy* (Lexington, Mass.: Heath, 1977).

[4] Although there has been fierce debate within the social science community concerning the actors in the contemporary world political system and whether their relative importance is shifting, most analysts would agree that the nation-state remains a primary actor. Thus, it makes sense to promote a research program that concerns national ocean policy. But nation-states are not the only actors on the world scene, and total involvement of a research program with the behavior of national actors may mask the important role played in ocean affairs by transnational public organizations, multinational profit-seeking organizations, transnational private organizations, and even individuals. Studies of national actors should be designed to take account of the input of these non-state influences on the national actors as well as on the international system.

coast, with varying degrees of authority, to 200 miles. Thus, although it is difficult to divide the ocean into parts, it is not impossible. However, the physical attribute of the ocean common still creates some policy problems for humankind even after the formal claims to jurisdiction are conceded. The oceans, like the air and other natural commons, are subject to multiple uses; some of them are possibly adverse to the owner's wishes. All users are affected by this condition when they attempt to manage the uses of the ocean. This is one factor in the argument that there should be national ocean policy.

The universality of the movement toward enclosure is an argument that national ocean policy actually does exist *de facto*. Because the 200-mile zones encompass most of the congested portions of most of the world's marine transportation routes, the most productive fisheries, all existing offshore oil and gas production facilities, and most marine recreation, there is little left in the oceans except polymetallic nodules and nuclear submarines (tactical and strategic) for the next generation of policy-makers to concern themselves about. The myriad problems being tackled by governments have an underlying similar basis. Comparative analysis might provide useful insights to those struggling with this class of decision.

A survey was conducted by me and some of my colleagues in order to assess the state of the art of academic national ocean policy studies.[5] The results of this survey served as the preparatory paper for the workshop. The survey dealt with works primarily in the English language published since 1967 that were concerned with the substance and process of national ocean policy. Eight research foci were chosen for special emphasis:

1. Organization and structure of national ocean policy
2. Ocean environment and coastal zone management
3. Ocean research and engineering development
4. Ocean national defense and policing
5. Fishing
6. Ocean minerals
7. Ocean energy
8. Ocean transportation, port and harbor management, and shipbuilding.

Coders examined approximately 6,000 items, of which 966 were thought to deal with national ocean policy. These were computer-coded.

[5] Robert L. Friedheim and Robert Bowen, "Assessing the State of the Art in National Ocean Policy Studies," Occasional Paper No. 1, Institute for Marine and Coastal Studies, University of Southern California, April 1978. A revised version has been published in *Ocean Development and International Law*, 7:3, Winter 1980, pp. 179-220.

The following type of information was recorded: country; marine region; United Nations regional or caucusing group; which part of the decision system the work dealt with (e.g., input, process, output); process (e.g., legislative, executive); central, regional, or local level; methodology used; and sector (e.g., ocean minerals, fishing). We also attempted to discover which works included policy recommendations or made attempts to forecast future conditions. Eight major general observations resulted from the systematic survey of the literature:

1. We know most about the output stage of ocean policy in all states surveyed. We know much less about input to ocean policy, and least about the process by which ocean policy is made.

2. Of the limited number of studies concerned with the decision process, most deal with bureaucratic decision. The literature on legislative and high-level executive decision is limited. There is virtually no literature on judicial decision. The overall trend is toward more process-related studies.

3. Some of the literature is cyclical. There was a sharp rise in ocean policy output studies associated with the 1973-74 convening of a United Nations Law of the Sea Conference, then a decline, and then a more recent rise. However, there has been a steady rise in input studies, and apparently a sharp rise in process studies (too small a sample size makes the observation less than definitive).

4. We know most about United States ocean policy. The next two most recently studied ocean policies are those of the USSR and Japan. Less is known about western European ocean policy, and the ocean policies of most of the third world countries have been studied very little.

5. Sixty-six percent of the literature surveyed concerns intra-political group ocean problems. We know most about North America, less about Western Europe, and much less about the ocean affairs of Latin America, Group of Seventy-Seven, Africa, and Oceania.

6. Thirty-one percent of the works surveyed dealt with specific marine regions. The following is the rank order of availability of work: North Pacific, North Atlantic, Caribbean, North Sea, Mediterranean, South Pacific, Arctic, Central Pacific Indian Ocean, and the Antarctic.

7. As to sectors, we know most about national ocean policy, organizational problems, fisheries, coastal zone management, pollution, ocean minerals, ocean-related constabulary, and security problems. We know much less about policy problems of ocean science, engineering, transportation, and ocean energy.

8. We discovered that a good proportion of the literature is policy-oriented. Thirty-six percent of the studies made policy recommendations. In addition, twenty-eight percent of the authors attempted to forecast the future, but about half of them did so with tools no more precise than legal, historical, or first-person experience modes of analysis.

Discussions at the Workshop

In some twelve hours of meetings, discussion among the participants ranged widely over the organizing paper, the substance of national policies, the system of states, subnational and transnational processes worth studying, the most appropriate organizing notions, methodologies that might yield fruitful results, and recommendations to the Marine Science Affairs Program. No consensus emerged on any of these subjects.

Despite the diversity of opinions expressed, what did emerge were a limited number of foci, basic thrusts, and organizational notions. I believe that they encompass a very representative range of intellectual foci that might be applied to the study of national ocean policy. While eight different foci or approaches are representative of the full range of opinion of the workshop participants, there was greater support (but no consensus) for one of them: that the worldwide enclosure movement is a phenomenon that inevitably must have an impact on virtually all national ocean policies. This opinion was shared by a substantial number of workshop participants. Indeed, two of the participants subsequently wrote a guest editorial for *Science* magazine pointing out the importance of the phenomenon for science, resource use, and environmental management.[6]

Enclosure

Many felt that enclosure could be looked upon as the primary independent variable in future studies of political decisions relating to the oceans.

[6] David A. Ross and Edward Miles, "The Importance of Marine Affairs," *Science* 201 (July 28, 1978):4353.

However, what is not clear is which dependent variables should be employed. But certainly among the most important are the implementation strategies being employed, the national variations in implementation strategies, the workability and efficiency of the attempts of virtually all nation-states to implement and enforce their enclosure decrees, and the short-, middle-, and long-run consequences to nation-states, to other participants in the international system, and to the international system itself.

Allocation of Scarce Resources

Several workshop participants viewed ocean policy as one more arena (albeit a wet one and a physical common) in which the foremost criteria for analysis is whether the ways humankind uses the resources of that arena meet a standard of efficiency. Scarcity of the resource means that we ask similar questions here as in any other arena where scarcity is a factor. According to some workshop participants, too often we have attempted to espouse goals in the ocean without reckoning the costs or attempting to measure the benefits in relation to the costs.

The Ocean Policy Component of the Regulation of Post-Industrial Society

The general structure of society in the post-industrial age has taxed scarce resources and fragile areas such as the oceans. The structure of the situation is the independent variable and ocean uses is the dependent variable. Proponents of this point of view see the task of ocean policy analysis as the development of sensible policies to avoid the evils of overexploitation. In short, national ocean policy studies should teach decision-makers to manage scarce resources in an era of limits. The purpose of policy and the studies which support them is to avoid the unintended consequences of profligate ocean use. Coastal zones are where the current major problems of ocean use occur and, therefore, studies on multiple-use conflicts found there should be emphasized.

Oceans as a Sector of National, Political, Economic, and Social Systems

Ocean policy concerns are an indicator that oceans are a subsector of national systems. That alone justifies the study of them. Some of the normal tools of political science can be employed if ocean policy is isolated for analytic purposes from the national system. Thus, country or area studies can be performed. Comparative analysis of countries, or within regions, showing similarities and differences in national or regional approaches to ocean use can provide useful insights. Scholars in this tradition can contribute years of specialized training regarding different cultures, values, ideologies, and languages that will help them provide insights that are not usually obtained through other approaches.

Economic Equity, or an Emphasis on the Distributional Aspects of Ocean Policy

Several workshop participants pointed to the fact that the rules of the previous international ocean system, which appeared to be equitable, have had inequitable consequences. The major ocean-using states, they said, have captured a disproportionate percentage of the economic value in using the ocean. National ocean policy studies should be viewed as an opportunity for correcting historic injustices by pointing out the policies that would lead to a just distribution of ocean resources in the future. Thus, ocean policy studies of developed states should emphasize the path they should follow that would lead to international equity; ocean policy studies for developing states would emphasize those strategies that would help them catch up and take their rightful places in the international political system.

Building an Ocean Policy from Scratch, or the Dilemma of the Developing Countries

A number of workshop participants pointed out that developing countries form a special case in national ocean policy studies. Although many developing states have ocean interests and make decisions affecting the ways in which their citizens and those of other states use the oceans, they rarely have an identifiable ocean policy. In a large number of cases, developing states must begin the process of forming an ocean policy from scratch. Because their situation may be substantially different from that of the developed countries, it may well be that the substance and process of their policies and the structure of their oceanic policy system will be quite different from that of the developed countries. Whether this is so, and if it indeed proves to be so, is a fruitful area of national ocean policy research. However, since physical parameters of the oceans constrain some ocean policy of the developed countries in the same way they do the developing countries, some lessons may be learned from the experience of the former. But in this latter case, the adaptations cannot be automatic, but must be carefully fitted to the situation in which the particular developing country finds itself.

Problem Solving

Despite the obvious limitations imposed by the academic schedule, some participants felt that university-based scholars should be asked to contribute to problem-solving-oriented research on national ocean policy. These participants were primarily concerned with "how to" projects. For them, it was sensible to ask purveyors of fundamental knowledge to apply what they knew to finding solutions to practical ocean policy problems. The following are typical of the research questions recommended: How can we restore the United States merchant fleet to the status it enjoyed in the era of the clipper ships? How best can we manage the United States 200-mile fishing conservation zone? How should the United States use its lead in ocean knowledge to gain an advantage in the future exploitation of the oceans? National decision-makers need this type of research if they are to make use of the ocean opportunities that are unfolding.

Inductive Approach

While emphasis in discussion was upon national ocean policy systems, some participants felt that something as large as national systems could be approached inductively, moving step by step, from the particular to the general, rather than merely dealing with the general *per se*. Microanalytic studies of the local and regional components of a national ocean policy community should provide insights not normally available if systemic studies are pursued exclusively. These could then be aggregated later into a meaningful picture of the whole. The types of studies suggested were analyses of the policy impact of new scientific knowledge or studies of the workings of the communities or organizations of ocean users such as fishermen.

Conclusions and Recommendations

Although the group that was assembled for the workshop was too heterogeneous to make specific recommendations, a considerable degree of agreement (but not consensus) was achieved. There was broad agreement concerning the idea that national ocean policy studies should be pursued, that numerous funding agencies should have a role in promoting national ocean policy studies, that any of several foci would be acceptable, and there was some agreement regarding considerations in evaluating national ocean policy proposals that relate primarily to enclosure. Let me now be more specific about the conclusions and recommendations of the workshop.

National Ocean Policy as a Focus

It was agreed that an important focus of ocean research should be the policies, decisions, and alternatives being pursued by contemporary nation-states regarding the use and management of ocean space. This should include asking how and why decisions are made as well as the evaluation of costs, benefits, and impacts of ocean decisions, particularly those that affect and are affected by new knowledge of the natural world.

Marine Science Affairs Program

It was recommended that the Marine Science Affairs Program finance studies in the area broadly identified as national ocean policy studies. These are valuable in and of themselves, but they also provide the contextual setting for studies of the direct impacts of new knowledge. There is no other funding agency within the United States government that is emphasizing such studies, and thus the Marine Science Affairs program is fulfilling an important need.

Because there is no realistic prospect that funding of the Marine Science Affairs Program will grow significantly, it is further recommended that other funding agencies representing both mission and non-mission organizations share the responsibility of funding national ocean policy studies. The subjects to be examined under the national ocean policy label touch very broadly on the responsibilities of virtually all government agencies. Previous studies have demonstrated that the proportion of study money spent to understand the social, political, and economic constraints on and consequences of using the oceans is well out of balance with what is spent to understand the physical constraints and opportunities of using the ocean. Until we better prepare ourselves by appropriate studies to understand the costs and consequences of what we might do to the oceans, we are subjecting government decision-makers to a greater probability of error than is necessary.

Purpose of National Ocean Policy Studies

Virtually all of the workshop participants agreed that the purpose of a program to sponsor national ocean policy studies is to improve the analytic base for advocacy. The intent of the program should be to provide the best policy analysis.

Foci

It was agreed that all eight foci described in the previous section, in the hands of skilled analysts, could provide useful analytic results. The variety of approaches reflect different perceptions of need for analysis concerning the ocean world as well as different types of analytic methods.

Alternate Funding Strategies

Although participants in the workshop were able to reduce to eight the variety of foci or approaches, we could not achieve consensus on the absolute primacy of one of these. I will report, in descending order of support at the workshop, four alternate funding strategies.

First, many felt that we should emphasize the causes, process, impacts, and consequences of the enclosure efforts by the present generation of nation-state decision-makers. This strategy was supported by a majority of participants. They claimed that the uses of the ocean for generations to come is being heavily influenced if not decided by the decisions relating to placing under national jurisdiction most activities in and under the waters of ocean areas up to 200 miles from coasts.

The second most popular suggestion for a thrust for a research program was to concentrate the studies on analyses—frequently suggested as comparative in method—of the major ocean-using states. This group of countries would include the United States, the USSR, Japan, Canada, the United Kingdom, France, Norway, and Brazil. These are the states that are actively involved in exploiting and degrading the oceans, and studies should focus upon matters such as the policy patterns they establish, the institutions they use and how they work, and the outcomes and consequences of their practices as they struggle to solve ocean problems in the succeeding generation. Thus, we can learn from a comparative study of successes and failures.

Some of the participants thought that the ocean problems of developing countries, their motives for making ocean policy, and the interests which they wish to see pursued could appropriately serve as the focus for meaningful studies. Developing countries are already playing an important role in multilateral ocean affairs and are likely to play a powerful and different role in enclosure. There was little disagreement among the workshop participants on the distinctiveness of ocean problems in the developing countries, but there was even less agreement on the comparative advantage to the Marine Science Affairs Program of placing special emphasis on the study of problems of developing countries.

Another research strategy might be to fill the "holes" in the existing literature identified in the survey of the state of the art. With the exception of ocean transportation where there may be a large, high-quality, specialized body of literature that was not identified in the survey, there was general agreement that the survey successfully identified areas where work was needed on input, process, output of national ocean decision systems, which states were not adequately covered, and which sectors (e.g., fishing, coastal zone management, etc.) were neglected.

Considerations in Evaluating National Ocean Policy Proposals that Relate to Enclosure

Since so much of the workshop was devoted to suggesting studies related to the causes, effects, and consequences of enclosure, more information that is useful for research managers emerged on this subject than on others. Below are listed some considerations for judging whether or not a proposed study of some aspect of enclosure would be worth pursuing. The list is cumulative, in that no proposed study need have all the listed attributes, but the study should possess attribute 1 as well as some of the others.

1. The proposed research has scientific rigor or analytic thoroughness in its design.

2. The proposal deals with a state that is actively pursuing enclosure rather than merely issuing enclosure decrees. The indicators of "actively pursuing enclosure" are implementation costs, enforcement problems, or adjustments, a visible politics of enclosure, and allegations of important consequences to new and traditional users of ocean space, domestic or foreign.

3. The project proposes to examine multiple-use problems in enclosure.

4. The project is designed to try to measure the impacts of enclosure, preferably with sophisticated tools, or by methods which at least allow comparison of the results of one study with another.

5. The project helps clarify the objectives that might be realized in enclosure.

6. The project deals with some aspect of enclosure that proposes to promote or retard the conduct of ocean science or help or hinder the effectiveness of international scientific cooperation.

7. Since the purpose of the research is to improve the analytic base for policy advocacy, the project should concern state, marine region, or policy problem of interest to policy-makers.

In an era in which governments have begun to focus on resource availability as a major political problem, more and more political decisions are being made on subjects on which our fundamental knowledge—social, political and economic—is scarce. It is essential to improve our knowledge of not only how natural systems work but also how human beings typically try to manipulate the natural world. If we are to deal intelligently with the oceans, we must move forward with vigor and perceptiveness with the task of increasing our knowledge. The workshop was a small step in this direction.